MIRACLE PROTEIN:

Secret of Natural
CELL-TISSUE REJUVENATION

Also by the Author

Health Secrets from the Orient

*Health Tonics, Elixirs and Potions for the
   Look and Feel of Youth*

Magic Enzymes: Key to Youth and Health

The Miracle of Organic Vitamins for Better Health

Natural and Folk Remedies

Natural Hormones: The Secret of Youthful Health

The Natural Laws of Healthful Living

MIRACLE PROTEIN:

Secret of Natural
CELL-TISSUE REJUVENATION

Carlson Wade

Foreword by William S. Keezer, M.D.

Parker Publishing Company, Inc.
West Nyack, N.Y.

Library of Congress Cataloging in Publication Data

Wade, Carlson.
  Miracle protein.

   1. Proteins in human nutrition. 2. Rejuvenation.
I. Title. [DNLM: 1. Dietary proteins--Popular works.
QU55 W119m]
QP551.W23   613.2'8   74-23200
ISBN 0-13-585653-1

Printed in the United States of America

DEDICATION

To YOUR Renewed Youth and Health.

FOREWORD
BY A DOCTOR OF MEDICINE

A miracle substance, found in certain everyday foods, holds the secret of extended youthful health. Feed this miracle substance to your billions of body cells and tissues, and you help unlock the mysteries of perpetual rejuvenation of your body and mind. This is the discovery made by Carlson Wade, a highly acclaimed nutritionist, writer-reporter in this priceless book.

The secret "youth factor" is known as miracle protein. We know for certain that protein does have a unique rejuvenation factor. It can help protect and relieve many symptoms of aging. We strongly believe that better nutrition with emphasis upon protein can work miracles in helping to slow down and even reverse the aging process.

This book pinpoints hundreds of conditions of aging, and shows specifically how you can use miracle protein to nourish your billions of cells and tissues to help correct these disorders. Carlson Wade has created one of the most brilliant books on the effective use of protein for extending youthfulness from head to toe, inside and outside.

The book is free of medical jargon. Yet, it is thoroughly researched and documented with convincing proof that miracle protein is the all-natural secret for cell-tissue rejuvenation and extending the prime of life.

Carlson Wade has prepared a youth-building compendium, with hundreds of easy-to-follow, step-by-step programs using protein to help you look and feel better. These programs help you enjoy a lengthy "age free" lifespan.

This outstanding nutritional researcher begins by offering proof that protein is the miracle food for the DNA-RNA "genetic codes" of your cellular system. Then he shows you exactly how you can easily use protein to feed these "codes." Once protein-ized, your body becomes transformed, and you radiate youthful energy, vitality and good health.

Carlson Wade shows you how to use protein to rejuvenate your skin, nourish your scalp for better hair growth (hair is 97 percent protein) and rebuild your brain cells as built-in protection against senility. He outlines easy-to-follow programs for rebuilding your internal organs with a variety of everyday cell-tissue feeding protein foods.

He offers documented proof that protein can help loosen up arthritic-like stiffness and enrich your bloodstream so it fairly "sings" with youthful vitality. He shows how to use protein to wash and clean your arteries and guard against excessive cholesterol buildup. He shows you the way to use protein to build youthful immunity against allergies, stress and tension. One powerful chapter shows you how to use protein for nourishment of your glands so they pour forth "forever young" hormones.

Carlson Wade then gives you little-known "miracle protein youth secrets" from the rejuvenation clinics and spas of Europe and the Orient. You are given natural protein formulas that reportedly give a "million dollar look and feel of youth."

All of these protein foods are available at almost any local market. Their cost is minimal. You probably have many in your own kitchen. Their benefits are priceless. With Carlson Wade's easy-to-follow programs, you can begin right now to supercharge your body with miracle protein, the secret of natural cell-tissue rejuvenation. Everything can be done right in your own home. No fuss. No bother.

Carlson Wade has prepared a highly recommended book. It is thoroughly documented and researched. It shows you how to use natural protein to slow down and reverse the aging process. This may well be the best book on rejuvenation that we have yet to see.

William S. Keezer, M.D.

WHAT THIS BOOK CAN
DO FOR YOU

This book takes the wraps off the mystery of "aging" and shows you how everyday protein foods can be used to nourish and replenish your billions of body cells and tissues to help create "new youth" of your body and mind.

There have been many professional reports about the miracle of protein but these have been prepared for the medical and scientific world and are rarely seen by the average reader. This book has drawn from the cream of the medical crop and reveals the secrets of food protein and how it can boost cellular replication and open the doorway to a look and feeling of "new youth" at any age. For the first time, here is a book written in simple, everyday language, offering the scientific discoveries in a step-by-step format, making it easy for you to follow, right in your own home.

This book offers a treasure of benefits for folks who have problems of aging skin, hormone inadequacy, weak circulation, joint aches, digestive distress and hundreds of other protein-deficient ailments. This book shows you how you can use everyday food proteins either singularly or in a combination to help renourish and regenerate the body cells and tissues so that these ailments healthfully can be soothed, eased and corrected. *This book shows you how science can make you look and feel younger with cell-feeding protein.*

If you are a man or woman living alone, you can use a variety of these clearly outlined protein programs to improve your appearance and general metabolism to enjoy better youth. The programs can easily be followed by just one person in a few moments of spare time.

If you are a homemaker or head of the family, you are eager to use protein to improve your youthful vitality and to help your family look and feel bouncy and alert with the knowledge of this book. It's surprisingly simple (delicious, too) to use these protein programs for yourself and your family so all of you will radiate the joy of cellular youth.

While many other books on protein tell you how important this miracle nutrient is for general health, the book you now hold is unique because it shows you how to use protein for specific cell-tissue rejuvenation programs. It shows you the simple *specific protein foods* that exert miracle rejuvenation upon the corresponding body parts. This book is especially beneficial for folks who want to correct *specific* ailments and want to know which protein foods can replenish the cells and tissues involved in the disorders. *This book lists these foods and tells you how to use them, when to use them, and what they will do to correct these ailments.* All this is prepared for your easy reference and convenience so you can use miracle protein with as little effort as possible for as much cell-tissue rejuvenation as Nature provides. This book will benefit everyone who wants clear-easy instructions on how to use miracle protein for overall top-to-bottom improved health, vitality and youthfulness.

A unique feature, not found in any other protein book, is the special PROTEIN PLANNER INDEX. If you have any special health condition, just consult the index to locate the protein food and the remedy to help you plan corrective rejuvenation. It offers you "at-a-glance" advice on a treasury of health ideas and how protein planning can help boost and extend the youthful prime of life. This *exclusive* PROTEIN PLANNER INDEX closes the gap between technical science and easy reading. It enables you to pinpoint the protein planning program for your needs in just a glance. The special index is a road map for the new youth waiting for you with miracle protein.

Everyday foods available at your corner market and at health stores (many are right in your pantry) can be your source of miracle protein, the stuff that feeds your cells and tissues and creates the chain reaction of body rebuilding and perpetual youth.

Enter a world of better health and vitality with this book as your guide. It may well prove to be the most valuable book on rejuvenation you may ever have to use. Begin the first step with the first chapter . . . .

CARLSON WADE

CONTENTS

5

1

MIRACLE PROTEIN—
THE NATURAL WAY TO
SELF-REJUVENATION

Nature has a secret "forever young" plan waiting for you. This secret plan will help you look and feel like a million dollars, as the joy of youth bubbles throughout your rejuvenated body from head to toe. It will help you enjoy continued vitality in your "forever young" extended lifespan. *This secret plan will help you stay young in body and mind even when the calendar shows you have had many birthdays.* This plan will be revealed in the pages of this book.

How Miracle Protein Protects Against Aging. Nature's secret "youth plan" calls for using miracle protein as a protective agent against aging. This substance, found in many foods, has a special function in giving you a "forever young" body and mind. It is food for your cells—and *the key to perpetual youth is in protein-ized cells.* The secret here is that your body is the end result of about 20 years of growth. You began life as the product of two cells. Eventually you developed over 100,000 billion cells. These cells determine your youthful health and lifespan. These cells depend upon one *basic* source of nourishment—*protein.* If you feed your cells "complete protein" and keep them protein-ized, you are feeding yourself Nature's "secret plan" for perpetual youth.

Secret of Protein-ization of Cellular Rejuvenation. Your cell is the basic unit of every single organism in your body and your mind. Your skin, muscles, organs, blood, bones, tissues, veins, arteries and just about every part of you is composed of cells. During the aging process, cells become worn out, divide and disintegrate, and new cells are then created through the natural biological process. *All of your body's basic biochemical processes begin or occur in your cells.* This process is not carried on by itself. Rather, the energy or the sparkplug that ignites this youth and life-giving process comes from one basic source—*protein*. This substance is the food, sustenance and very life of your cells. And your body is the end result of protein-ized cells.

The secret of youth is to have enough protein to create the formation of more cells than those that die off in the natural biological process. Protein protects against the age-causing process of inadequate cell formation. *Protein helps keep you young by balancing the amount of dying cells with brand new young cells.*

HOW PROTEIN OFFERS "TRIPLE REJUVENATION" FOR YOUR CELLS

While cells may be diverse, they have *three components* which need protein for perpetual youth. When protein is available, it can offer "triple rejuvenation" in this manner:

1. *Cell Membrane.* The "wall" or envelope in which the cell is contained. This "wall" needs adequate protein to remain strong and resilient, and to resist the ravages of daily living. Strongly built protein-nourished cell membranes become a fortress against aging.

2. *Endoplasmic Reticulum.* A substance within the cell which needs protein for nourishment, and which also serves for the exchange of metabolites.

3. *Cytoplasm.* (cyt=cell, plasm=matrix or something molded), formerly considered protoplasm, but now known as the source of the cell's biological protein activities.

Let us think of the cell as an industrial unit within the body. The nucleus is the office, and the cytoplasm is the factory. The energy or force that keeps the office and factory going is that of *protein*. Shut off protein and the cells start to die, and the body undergoes aging. Protein is the secret plan for smooth body function and efficiency.

HOW PROTEIN NOURISHES THE DNA-RNA FACTORS— KEYS TO ETERNAL YOUTH

Two factors located in and around your cells depend upon a never-ending supply of protein to keep you alive and young. These are known as:

1. DNA. (Deoxyribo-nucleic acid.) Considered a "central director" of all youth-creating processes, DNA consists of various molecules that are used to create new cells during the natural birth-death-rebirth process of internal repair. When a new cell has to

form, DNA molecules uncoil and divide in the center, and each half of the DNA replicates its other half again. Thus, there are two new DNAs which become the nucleus of a new cell. The original cell has divided and shaped into two brand new DNAs. This rejuvenation-cell factor is energized by protein, which is food for DNA. No protein, or a lack of adequate protein causes the DNA molecule to become weak, sluggish and inadequate. Cellular rejuvenation slows down or halts abruptly. This results in the aging process. But with sufficient protein, the DNA molecule, confined in the nucleus, directs the creation of rejuvenating cells throughout your body from head to toe.

2. RNA. (Ribo-nucleic acid.) A second kind of nucleic acid, or substance within the cell, is RNA, which is a single-strand helix or spiral coil. It acts as your cell's messenger. RNA duplicates the activities on the DNA and then transfers them to the area in the cytoplasm where the cell manufactures its proteins. RNA is a mold for the synthesis of life-giving protein. The properties of a protein depend on the sequence of amino acids, which is determined by the sequence of links in the RNA chain, or ultimately by the kind of DNA in the nucleus of your cell.

Therefore, the self-perpetuation of your cells depends upon a vigorously functioning DNA and RNA substance. These two substances depend upon protein for energy and life force. A well-proteinized DNA and RNA substance means youthful cells and the hope for perpetual rejuvenation.

Cells Hold Secret of Youth. The billions of body cells are constantly being repaired and replaced, as DNA is used to divide and subdivide them to create internal "rebirth." Every cell includes a replica of the DNA. *Protein* is the key or the "secret" force that energizes and nourishes the DNA and RNA substances to create this life-giving process.

After DNA sends forth its directions, RNA acts as a messenger to form the new protein-built cell. DNA gives protein to RNA, which uses it for the synthesis of new cells. At all times, protein is needed for this cellular rebirth. It holds the secret of youth.

HOW CELLULAR REBIRTH HELPS YOU LIVE YOUNG LONGER

You may wonder about being concerned about protein nourishment, when the automatic body process is to form new cells when old ones die. But there is a little-known reason that makes protein a vital life-giving substance. *Rebuilt cells may not be as strong as the original ones if protein is deficient in the system.* Here are three reasons:

1. Cells that have gone through a number of divisions are less efficient than they were in their original state.

2. There are a limited amount of cell divisions available. Cells do *not* permanently keep dividing and rebuilding. In some reported tests, cells were able to divide some 50 times, and then died—*without being replaced.*

3. Cellular division and rebuilding is more vigorous in the age bracket of birth and 20 years. Over the age of 20, cellular rejuvenation is reduced. More cells wear out and die than are rebuilt. This leads to the aging process.

Protein: Cellular Elixir of Youth. The more protein available for cellular rejuvenation, the greater is the hope of perpetual youth. Protein is food for the cells and the elixir of youth.

Each day, millions of cells die in your bodies. True, some of them are replaced, but many are not because of inadequate protein. For those cells that are not replaced, the aging rate is slow, but is noticeable. If a cell splits wrong, it may become useless; worse, it may become malignant and cause illness. But Nature's secret plan calls for using "protective protein" as a cellular elixir of youth!

Science has discovered that all living matter, in its simplest form, is *protoplasm*—the substance which composes the "living stuff" of the cells of your body from top to bottom. Protoplasm is nourished by protein, which is the hope for eternal youth through cellular rejuvenation. Now, let's see how your life (and youth) depend upon protein.

OFFICIAL U.S. GOVERNMENT STATEMENT ON YOUTH-BUILDING PROTEIN

Dr. Ruth M. Leverton, in the official government report, *Food,*[1] has this to say about youth-building protein:

"You are looking at a superb package of proteins when you see yourself in a mirror. All that shows—muscles, skin, hair, nails, eyes—are protein tissues. Teeth contain a little protein.

[1] Leverton, Ruth M.: *Food, The Yearbook of Agriculture,* United States Department of Agriculture, Washington, D.C., 1959, pp. 57-70.

"Most of what you do not see is protein, too—blood and lymph, heart and lungs, tendons and ligaments, brain and nerves, and all the rest of you.

"Genes, those mysterious controllers of heredity, are a particular kind of protein. Hormones, the chemical regulators of body processes, and enzymes, the spark plugs of chemical reactions, also are proteins. *Life requires protein.*"

Most Plentiful. "Next to water, protein is the most plentiful substance in the body. If all the water were squeezed out of you, about half your dry weight would be protein. About a third of the protein is in the muscle. About a fifth is in the bone and cartilage. About a tenth is in the skin. The rest is in the other tissues and body fluids. Bile and urine are the only fluids that do not contain protein."

Blood Protein. "There are several dozen proteins in the blood alone. One of the busiest is hemoglobin, which constantly transports oxygen from the lungs to the tissues and brings carbon dioxide back from the tissues in the lungs. Ninety-five percent of the hemoglobin molecule is protein. The other 5 percent is the portion that contains the iron."

Natural Immunity. "Other proteins in the blood are defenders," says this official U.S. Government authority, "for they give us the means of developing resistance and sometimes immunity to disease. Gamma globulin can also form antibodies, substances that can neutralize bacteria and viruses and other micro-organisms. Different antibodies are specific for different diseases. *The presence of an antibody in the blood (created by protein) may give the person immunity to the disease.* At least it gives him a head start in fighting the virus, and the disease will be less severe. Gamma globulin also helps the scavenger cells, or phagocytes, engulf disease microbes."

Maintain Exchange Balance. "Proteins help in the exchange of nutrients between cells and the inter-cellular fluids, and between tissues and blood and lymph. *When one has too little protein, the fluid balance of the body is upset, so that the tissues hold abnormal amounts of liquid and become swollen.*"

Promote Processes of Daily Living. "The proteins in the body tissues are not there as fixed, unchanging substances deposited for a lifetime of use. They are in a constant state of exchange. Some molecules or parts of molecules are always breaking down, and others are being built as replacements."

"This exchange is a basic characteristic of living things; in the body it is referred to as the dynamic state of body constituents—the opposite of a static or fixed state. This constant turnover explains why our diet must supply adequate protein daily, even when we no longer need it for growth. The turnover of protein is faster within the cells of a tissue (intracellular) than in the substance between the cells (intercellular)."

Protein Produces Energy. "Proteins, like starches, sugars and fats, can supply energy. The body puts its need for energy above every other need. It will ignore the special functions of protein if it needs energy and no other source is available. This applies to protein coming into the body in foods and to protein being withdrawn from the tissues. Either kind gets whisked through the liver to rid it of its nitrogen (a basic element needed for youthful living, and a component of protein) and then is oxidized for energy without having a chance to do any of the jobs it is especially designed to do. The protein-sparing action of carbohydrates means that starches and sugars, by supplying energy, conserve protein for its special functions." This is the official U.S. Government report on the youth-building powers of cell-rejuvenating protein.

HOW MUCH PROTEIN DO YOU NEED?

The official Food and Nutrition Board of the National Research Council has set a recommended *daily* allowance as follows:

Adult males, average height and weight . . . . 70 grams daily.
Adult females, average height and weight . . . . 60 grams daily.

Special Note: These are the basic *minimums* needed for daily survival. For better health, the food program should have a bit more protein to give the body its needed working materials for the building and rebuilding of billions of body cells. Also, the stresses and strains of daily living and the middle years when cellular reproduction has slowed down, call for increased protein intake. But—the daily food program should contain a *balance* of vitamins, minerals, enzymes, essential fatty acids, carbohydrates and calories since the key to youthful health is in *internal harmony*. No one food should dominate another since this disturbs the cellular replenishment activities of protein. Your goal should be adequate protein with a balance of other nutrients for better health of body and mind and youth extension.

Where can you obtain protein? Consult the Chart 1 shown here for the official counts for protein planning, as drawn from official U.S. Government listings.

PROTEIN CONTENT OF FOODS—OFFICIAL
U.S. GOVERNMENT LISTINGS[2]

FOOD	AMOUNT	GRAMS PROTEIN

Milk and Milk Products

Milk, Cow:		
Whole or non-fat fluid . . . . 1 cup		8.5

Chart 1

_____
[2]Leverton, Ruth M.: *Food, The Yearbook of Agriculture,* United States Department of Agriculture, Washington, D.C., 1959, pp. 71-73.

FOOD	AMOUNT	GRAMS PROTEIN
Non-fat dry, instant . . . . 1 tablespoon		1.8
Buttermilk, cultured . . . . 1 cup.		8.7
Milk, goat. . . . . 1 cup.		8.1
Milk, human . . . . 1 cup.		3.4
Cheese:		
Cheddar . . . . 2 tablespoons		7.1
Cheddar, processed . . . . 2 tablespoons		6.6
Cottage . . . . 2 tablespoons		4.8
Cream . . . . 2 tablespoons		2.6

Eggs

Whole, large . . . . 1 egg		6.4

Meat, Poultry and Fish, Raw

Beef, medium fat, without bone . . . . 1/2 cup		20.6
Chicken, fryer, flesh only . . . . 1/2 cup		23.4
Fish. . . . . 1/2 cup		20.6
Heart, beef . . . . 1/2 cup		19.2
Kidney, beef. 1/2 cup		17.0
Lamb, leg, without bone . . . . 1/2 cup		20.4
Liver, beef . . . . 1/2 cup		22.3
Sausage:		
Bologna . . . . 2 tablespoons		4.2
Frankfurter . . . . 1/10 lb.		6.4
Turkey, flesh only . . . . 1/2 cup		27.2
Veal, round, boneless . . . . 1/2 cup		22.1

Mature Legumes and Their Products

Beans, common . . . . 2 tablespoons		6.1
Chickpeas, garbanzo . . . . 2 tablespoons		5.9
Cowpeas . . . . 2 tablespoons		6.5
Lentils. . . . . 2 tablespoons		7.1
Lima beans . . . . 2 tablespoons		5.9
Peas. . . . . 2 tablespoons		6.7
Soybeans . . . . 2 tablespoons		9.9
Soybean flour, low fat . . . . 1 cup.		45.1
Soybean milk . . . . 1/2 cup		3.9

Chart 1 (Continued)

FOOD	AMOUNT	GRAMS PROTEIN

Seeds, Nuts and Their Products

FOOD	AMOUNT	GRAMS PROTEIN
Brazil nuts	2 tablespoons	4.1
Coconut, fresh	2 tablespoons	1.0
Cottonseed flour	2 tablespoons	12.0
Filberts	2 tablespoons	3.6
Peanuts	2 tablespoons	7.6
Peanut butter	1 tablespoon	4.2
Pecans	2 tablespoons	2.7
Sesame meal	2 tablespoons	9.5
Sunflower meal	2 tablespoons	11.2

Grains and Their Products

FOOD	AMOUNT	GRAMS PROTEIN
Barley	2 tablespoons	3.6
Bread (4% non-fat dry milk, flour basis)	1/20 lb or 1 slice	1.9
Buckwheat flour, dark	1 cup	11.5
Corn and soy grits	1 cup	9.0
Corn Products:		
Flakes	1 cup	2.0
Grits	1 cup	13.9
Meal, whole	1 cup	10.9
Oatmeal	1 cup	11.4
Pearlmillet	2 tablespoons	3.2
Rice, white	1 cup	14.5
Rye flour, light	1 cup	7.5
Sorghum, grain	2 tablespoons	3.1
Wheat Products:		
Flakes	1 cup	3.8
Flour, whole grain	1 cup	16.0
Flour, white	1 cup	11.6
Germ	1 cup	17.1
Macaroni, elbow	1 cup	15.7
Noodles	1 cup	9.2
Shredded wheat	2 tablespoons	2.9

Vegetables, Raw

FOOD	AMOUNT	GRAMS PROTEIN
Beans, lima	4 tablespoons	4.3
Beans, snap	4 tablespoons	1.4
Cabbage	4 tablespoons	.8

Chart 1 (Continued)

FOOD	AMOUNT	GRAMS PROTEIN
Carrots	4 tablespoons	.7
Collards	4 tablespoons	2.2
Corn, sweet	4 tablespoons	2.1
Cowpeas	4 tablespoons	5.3
Kale	4 tablespoons	2.2
Okra	4 tablespoons	1.0
Peas, green	4 tablespoons	3.8
Potatoes	4 tablespoons	1.1
Spinach	4 tablespoons	1.3
Sweetpotatoes	4 tablespoons	1.0
Turnip greens	4 tablespoons	1.6

Miscellaneous

Gelatin	1 tablespoon	8.6
Yeast:		
Compressed	2 tablespoons	3.0
Brewer's, dried	1 tablespoon	3.0

Chart 1 (Continued)

HOW ENZYMES TRANSFORM PROTEIN INTO YOUTH-BUILDING AMINO ACIDS

Food protein cannot be used by your body unless it is first transformed into *amino acids* by enzymes. (These are protein-like transformers in your body that act as catalysts. That is, they break down the protein in food so it can be turned into amino acids and then absorbed into the bloodstream to perform their cellular rejuvenation tasks.) Enzymes are found in *raw foods* such as raw fruits, vegetables, seeds, nuts, grains. . . and raw fruit and vegetable juices. Without these raw food enzymes, food protein would remain in a "lump" and be useless.

Secret of Super-Protein Power: Until she discovered the "secret" of enzyme-protein combinations, Alice B. began to look her age. Only 47, the housewife was always tired, had cold hands and feet, developed skin wrinkles, was constantly catching colds, even in warm weather. She insisted she ate more than her share of protein. But a look-see into her food program showed that nearly all of her meals were *cooked*. Almost nothing was *raw*. This was the error. She was told to make this amazingly simple but amazingly powerful correction:

Begin each and every protein meal with a large raw fruit or vegetable salad. End each and every protein meal with a large raw fruit or vegetable salad . . . or a raw juice.

Benefits: The enzymes in the raw foods then performed a process of *catabolism* upon the cooked protein foods. This process broke down the protein into amino acid compounds which could then be used by the body for cellular rejuvenation.

Results: Within 7 days of this easy enzyme-protein combination, Alice B. responded with youthful vigor. She felt warm in hands and feet, her skin was firmed up, and she could enjoy natural immunity to colds and most simple allergies. Now life was more than joyful, it was youthful again, thanks to the easy raw food program that put super-power into protein metabolism. Alice B. wouldn't think of any meal that was without raw fruits or vegetable juices. It was her "secret" of super-protein power.

HOW AMINO ACIDS CAN UNLOCK THE DOORWAY TO THE ROAD TO YOUTH

All proteins consist of very large molecules, which in turn are made up of various combinations of smaller building blocks, called amino acids. These include two groups:

1. *Non-essential amino acids.* While they are essential, they are so named because the body can prepare them out of substances already within the system. There are about 12 of these non-essential amino acids. But a serious protein deficiency means that they, too, may be inadequate, so do not take them too much for granted.

2. *Essential amino acids.* These have essential cell-building properties and the body is unable to synthesize them from simpler substances. They are essential because they must be supplied *pre -formed,* in the foods listed in this chapter. There are 10 of these essential amino acids. They can unlock the doorway to the road to youth.

Let's take a look at these 10 essential amino acids and see how they can help build better health, prolong life, rejuvenate your cells and offer you hope for perpetual youth through self-renewal:

1. *VALINE.* This amino acid is needed to spark mental vigor, muscular coordination, smooth emotional health. A deficiency may lead to nervous ailments, fingernail biting, poor mental health and insomnia.

2. *LYSINE.* Most body growth factors are dependent upon lysine. Blood is nourished with this amino acid, antibodies are formed. Visual disorders such as "red spider webs" in the eyes may be traced to poor lysine intake. Chronic tiredness and fatigue is another possible symptom of a deficiency.

3. *TRYPTOPHANE.* For rich, red blood, get this amino acid. For youthful skin and healthy hair, you need it. Since it helps in the utilization of the B-complex vitamin group, it also aids in better digestive powers. Nerves also need this amino acid.

4. *METHIONINE.* Liver and kidney cells need this amino acid for regeneration. Rheumatic ailments may also be relieved by adequate methionine nourishment. Influences the hair follicles and stimulates better hair growth. Also functions to remove poisonous wastes (detoxifying) from the liver, and protect against conditions of liver necrosis—destruction of the delicate tissues and capillaries of this vital body organ.

5. *CYSTINE.* This amino acid supplies over 10 percent of the insulin needed by your pancreas for metabolization of sugars and starches. Hair and skin health are also influenced by cystine.

6. *PHENYLALINE.* Your thyroid glands demand this amino acid for stimulation so they can secrete the iodine-rich hormone of thyroxine. This hormone sparks nerve activity and helps maintain mental balance. It also influences blood vessel health. Some visual disorders may be traced to a deficiency.

7. *ARGININE.* Males especially need this amino acid since seminal fluids contain as high as 80 percent of this protein by-product. A severe deficiency may cause sexual disorders and/or sterility. Another valuable function of arginine is to detoxify poisonous wastes and filter out toxic debris.

8. *GLUTAMIC ACID.* Brain health responds to this amino acid. In some situations, the intelligence and personality of the individual, together with overall mental and physical equilibrium have responded remarkably to this amino acid. Some of the most stubborn and erratic individuals have responded with youthful alertness when a deficiency of this amino acid was corrected.

9. *HISTIDINE.* The auditory nerve needed for good hearing is stimulated and "fed" by this amino acid. Deafness and hard of hearing problems have been relieved by corrective supplementation. The nerve cells of the hearing mechanisms need this valuable amino acid.

10. *THREONINE.* The digestive and intestinal tracts will function more smoothly with this amino acid. It appears to perform better metabolism and assimilation, too.

Each Amino Acid Is Important. There are several dozen amino acids. Yet, each has a separate function . . . but are all linked together to help build cells and tissues and promote a feeling of youthfulness. *Each amino acid does have a specific purpose of its own. One cannot be substituted for another. You need ALL amino acids for good health.*

FOUR VITAL YOUTH-BUILDING BENEFITS OF AMINO ACIDS

When you eat a protein food that is then broken down by raw foods enzymes into *usable amino acids,* here are the four basic youth-building benefits for your body:

1. Amino acids are building blocks which are used to make billions of body tissues and cells. These "blocks" are the raw materials required to synthesize growth and repair in *all* parts of your body.

2. Thousands of enzymes needed for food digestion are created by these amino acids. These wonder workers also help to produce hormones and glandular secretions. Reproductive powers as well as the strength of virility and fertility may be traced to an adequate amino acid supply.

3. Because amino acids have a colloidal osmotic pressure (energy and liquid balance), they are needed in the bloodstream to keep it moving along swiftly, flowing through your arteries and veins.

4. Amino acids produce youthful vitality. As soon as the specific *amine* has been chopped off from the amino acid, the molecule that remains has a lot of carbon but little nitrogen. If this molecular amino acid is not needed for immediate use, the body changes it into glucose and glycogen to be stored in your liver. It is then used in situations requiring energy. *You can give yourself "energy insurance" in the form of stored up amino acids.*

Remember this important secret: amino acids or proteins are NOT available from the air, from the water or from the sunshine. Your body *cannot* manufacture the essential amino acids. You can get them *from foods,* such as recommended by the U.S. Government in the charts in this chapter. Eat protein foods daily, together with other valuable nutrients in a balanced diet, and eat raw foods with each meal, and you'll be giving your billions of cells the working materials out of which "new life" can be created, helping to give you at least 50 health-packed years of youth . . . starting in your middle years!

HIGHLIGHTS

1. The secret of self-rejuvenation is in protein-ized cells and tissues throughout your body. Protein is food for your cells and youth for your body.

2. Protein offers "triple rejuvenation" to your cells and hope for a prolonged and extended prime of life.

3. Protein sparks the DNA-RNA factors which are keys to a feeling of eternal youth.

4. Protein is a recognized "elixir of youth," as described by official U.S. Government authorities.

5. Plan your protein intake, daily, the easy way, with the simple chart, taken from official U.S. Government listings. A variety of protein foods is needed for cellular rebuilding.

6. Alice B. combined raw foods with protein and sent a shower of youth-building amino acids to her cells, and she bounced back with a "second youth."

7. Amino acids can unlock the doorway to the road to youth. All are waiting for your better, renewed health in protein foods.

2

HOW PLANT AND OCEAN PROTEINS HELP GIVE YOU "FOREVER YOUNG" SKIN

Proteins from plant and ocean foods are prime sources of unique ingredients that can help give you an all-natural "face lift" so that you will enjoy the smooth texture of a "forever young" skin. The secret ingredient in these foods is *collagen,* the substance which can actually help smooth out wrinkles, reverse the aging process and put a "schoolgirl" glow on your face. This substance found in plant and ocean foods offers a miracle of rejuvenation for your skin—the largest organ of your body and the first to benefit from cellular rebuilding.

COLLAGEN—THE MIRACLE NATURAL "FACE LIFTING" PROTEIN

Collagen is the supportive protein of skin, muscle, bone and tissue. It is the all-pervasive ingredient that makes up to one-third of all the protein in your body. It is the firming protein in your skin, the matrix in your bones, the durable material of your body. Collagen is the "envelope" in which all of your organs are protectively wrapped; it is the cellular wall between your cells.

Collagen Keeps Skin Looking Young. Collagen is a protein that is the principal component of connective tissue. It firms up and smooths out furrowed, wrinkled skin, and it gives a look and feel of firm, smooth, resilient skin. A sufficient supply of collagen can help give the skin the appearance of having been "face lifted" with a "forever young" glow of youth.

Plant Protein Protects Against Collagen Deficiency. A powerhouse of unique protein is found in *Brewer's yeast,* a plant food that has been seen to be a prime source of nucleic acid, the substance that leads to the formation of collagen, the skin-youth food. This plant

30

food (sold at health stores and specialty shops as well as many supermarkets) protects against collagen deficiency and helps slow down and reverse the aging process.

Stimulates Formation of DNA-RNA Cellular Substances. Brewer's yeast contains a form of protein that helps stimulate the DNA-RNA cellular substances needed to use collagen for tissue rebuilding. Brewer's yeast is a prime source of collagen, needed for skin rejuvenation. In daily living, the body cells wear out and are replaced about every three months. The aging process takes place when the new cell does not reproduce exactly like the old cell it replaces because of a collagen deficiency. This is called "cross-linkage" whereby there is a "shrinkage" of the collagen molecules and fibrous proteins that support bone, tendon, connective tissues and cartilage. This "cross-linkage" strangling of the healthy cells leads to the aging of the skin.

Brewer's yeast is a prime source of *collagen,* the protein that reportedly is able to deactivate or counteract the effects of this "shrinkage" and to help "unlock" the links so that there is a free and regular exchange of nutrients to maintain youthful resiliency.

Collagen then helps create the formation and distribution of DNA-RNA, which are needed to help correct or reverse the gradual cell change or deterioration. This offers a miracle of self-rejuvenation of the skin.

Yeast Protein Boosts Cell Foods. The protein in Brewer's yeast helps create collagen, which then takes DNA-RNA for use to nourish the cells to help boost the body's sluggish production. (These cell foods are normally supplied by the body, but in lesser and lesser amounts as aging progresses.) Brewer's yeast helps take up the slowdown created during aging. It helps feed the body cells, helps form new cells that are identical to the cells they replace, and consequently creates *younger cells* which create a younger-looking body!

HOW MIRACLE PROTEIN HELPED REVERSE THE AGING PROCESS

A doctor reportedly has helped "rejuvenate" test patients in a program calling for daily intake of Brewer's yeast as the source of collagen for the formation of DNA-RNA cell foods. Benjamin Frank, M.D., reports[1] giving a set of patients a special protein program:

[1]*Nucleic Acid Therapy in Aging and Degenerative Tissues,* Benjamin Frank, M.D., Psychological Library, New York, 1969.

Youth Foods: Daily, a supplement to provide 30 milligrams of nucleic acid. (About 200 milligrams of Brewer's yeast would supply this amount.) Desiccated liver, large amounts of sardines, sweet-breads and a supplement of the B-complex vitamins and minerals.

Immediate Results: Speedily, the people displayed more energy and well-being. The most striking benefits appeared on the skin. There was no long wait. Results were almost overnight!

Wrinkles Vanished! Skin Smoothed Out! The doctor reported that the wrinkles decreased on the Brewer's yeast program; there was a tightening of the skin and an increase in natural moisture. Skin roughness was diminished. Skin of the back of the hand improved in smoothness in the vast majority of the cases.

Corrects So-Called Aging Ailments. The collagen and other protein in Brewer's yeast helped boost a healthful DNA-RNA function, so that this program was beneficial in reversing the so-called ailments of the aging. Patients who had coronary heart disease and congestive heart failure were improved. Heart function, itself, was clearly improved.

Improves Mental Health. Most of the patients responded with rejuvenated thinking. The *benefit* of collagen feeding here is that it appears to boost cerebral circulation. It protects against cellular degeneration of the brain neurons and associated structures. Collagen is the protein that helps boost cerebro-vascular circulation, sending a stream of nutrition to the brain so it can function more youthfully.

According to Dr. Frank, Brewer's yeast and a healthful protein program offers food for the cells of the body (and your skin is the largest organ of your body, the heaviest concentration of cells, and basically protein) and helps boost a form of self-rejuvenation. Your skin cells contain the blueprints for the formation of your body and need collagen as nourishment. As the supportive protein of your body, collagen offers hope for "forever young" skin.

HOW TO USE PLANT PROTEINS FOR SKIN YOUTH

Collagen is found in such plant foods as Brewer's yeast, nuts, whole grain cereals, peanuts, sunflower seeds, dried peas, wheat germ, sesame seeds, olives, avocados. These are prime sources of collagen, the needed cell-nourishing protein. To help boost the function of collagen, it is wise to *combine* the food with Vitamin C, since this acts as an energizer. It is Vitamin C that promotes the vitality for collagen so it can then build the DNA-RNA cell walls for a youthful looking, wrinkleless skin. A *combination* of these plant proteins with Vitamin C is the key to youthful skin rebuilding.

Official U.S. Government Recommendations for Vitamin C Foods. These foods, listed in *Family Fare*[2], are "Citrus fruits—oranges, grapefruit, tangerines, lemons and their juices, and fresh strawberries. Other important sources include tomatoes and tomato juice, broccoli, brussels sprouts, cabbage, green peppers, some dark-green leafy vegetables such as collards, kale, mustard greens, turnip greens, potatoes and sweet potatoes, especially when cooked in the jacket."

10 MIRACLE PROTEIN SKIN FOOD COMBINATIONS

Here is a variety of protein-Vitamin C food combinations that interact so that they offer you top notch collagen as well as needed Vitamin C to promote a healthful skin rebuilding from within your body:

1. Grapefruit wedges with diced nuts.
2. Tomato juice with desiccated liver and Brewer's yeast as a Skin Tonic.
3. Fresh strawberries with orange wedges and sunflower seeds, sprinkled with Brewer's yeast.
4. Steamed green peppers, green leafy vegetables, a half-teaspoon of Brewer's yeast flakes.
5. Fruit juice with a half teaspoon of Brewer's yeast flakes as a morning energizer.
6. Raw tomato and lettuce salad sprinkled with diced seeds, nuts, and sliced olives.
7. Whole grain cereal in milk, sprinkled with sesame seeds and wheat germ. Fruit slices.
8. A cup of peanuts (giving you over 50 grams of protein) together with a fresh fruit salad.
9. Fresh salad of avocado slices with green peppers, some cooked vegetables, and a side dish of sunflower seeds.
10. Baked potato in the jacket; cut open, sprinkle with Brewer's yeast and assorted seeds. Eat with a fresh fruit salad.

Benefits: These are high-protein plant foods that are your prime sources of skin-cell feeding collagen. But when *combined* with the

---

[2]*Family Fare: A Guide to Good Nutrition,* U.S. Department of Agriculture, Washington, D.C., 1970.

Vitamin C foods listed, the collagen is *energized* and *activated* to perform speedy regeneration of your skin cells, and help smooth out furrows and wrinkles, help "plump up" sagging skin and give you the look and glow of youthfulness.

HOW GLADYS G. USED PROTEIN FOR A NATURAL "FACE LIFT"

Gladys G. was beginning to look her age. Her skin was blotchy, sallow. She developed crease lines that grew deeper and deeper, until her face was wrinkled furrows. She felt chronic tiredness. The least bit of work made her exhausted. She began to feel as aged as she looked. She had to give up her part-time job at a local shop because she could not cope with even the simplest of tasks. Someone jokingly suggested she should get a "face lift," which hurt her deeply. But it made her take a good look at herself. She took stock of her ways of life. She set about to correct her protein intake and see if she could get a "natural face lift."

Simple Program

Breakfast called for a protein food such as a soft boiled egg, whole grain cereal, cottage or farmer cheese, a dish of seeds and nuts, and a raw fruit or raw fruit salad.

Luncheon was a *Protein Punch* of a raw vegetable juice in which she stirred one-half teaspoon of Brewer's yeast and added a half teaspoon of desiccated liver.

Dinner was a slice of lean meat (beef or poultry), a large raw vegetable salad sprinkled with Brewer's yeast, a dish of seeds and nuts, and a raw fruit juice.

Nightcap was tomato juice with a half teaspoon of Brewer's yeast stirred vigorously. A bit of lemon juice added zest as well as more Vitamin C to whip up the collagen so it could work while Gladys G. slept.

Three Weeks to New Youth. Gladys G. followed this simple protein program for three weeks. She avoided processed, artificial, synthetic and chemicalized foods. The protein-ization of her body, sparked by Vitamin C, boosted her previously deficient amount of collagen and now DNA-RNA went to speedy work to rebuild her tired, broken and aging cells. *Results?* After a three week protein program, she returned to her job looking so youthful, with her skin smooth as a youngster, her eyes alert, all wrinkles gone, and an exuberant vitality, that co-workers could not believe it was a "natural face-lift." Thanks to protein, Gladys G. looks ten years younger—and feels twenty years younger.

WHY OCEAN PROTEIN IS A RICH TREASURE OF SKIN PROTEIN

For optimum skin rebuilding, you need a *complete* protein food. Most foods contain protein in an *incomplete* balance, which means that not all amino acids are present. But Nature has given us *one* food that is a *complete* protein. Nature has put it in its oceans. That food is *fish*. It is one of the best protein foods for the skin, and the body, too.

Official U.S. Government Recommendations. Andrew W. Anderson, of the U.S. Bureau of Commercial Fisheries, in *Food*[3], explains, "Fish is about 18 percent protein, which is *complete,* well-balanced and not easily affected by the usual cooking methods. It is 85 to 95 percent digestible. *An average serving furnishes more than enough animal protein to meet the usual daily need for protein."*

Fish Protein: Miracle of Health. The protein in fish contains *all* of the biologically essential amino acids, the components necessary to your body for building and rejuvenating all forms of bodily tissues. These 10 essential amino acids, which are contained in good quantity in all fish protein, cannot be synthesized by your body and must be obtained from the diet.

The terms "biological value," "biologically complete" and "biologically balanced" all refer, when applied to a protein, to a measure of the nutritional value of that protein in terms of amino acid content.

Why Fish Protein Is a Prime Skin Rejuvenator. The protein in fish contains little or no cartilaginous material. This means it is quickly broken up by the digestive processes of the stomach and intestines and easily absorbed from the intestines into your body. These protein fragments, the amino acids, then are easily available to build and rebuild your body tissues. The collagen formed by fish protein is of a biologically balanced complete value so it can quickly form

---

[3] Andrew W. Anderson: *Food, The Yearbook of Agriculture,* United States Department of Agriculture, Washington D.C., 1959, page 353.

DNA-RNA needed for cellular regeneration. Fish protein is of high quality, it is *complete*, and it can speedily be used for quick rejuvenation. It is a highly desirable "youth" food.

HOW MICHAEL A. TURNED BACK THE "SKIN AGING CLOCK" WITH TWO MIRACLE FOODS

As a travelling salesman, Michael A. admittedly neglected eating the right foods. He had to subsist on packaged and processed foods. Overcooking and poor quality eating on the road soon gave him deficiency symptoms.

Michael A. had scaly skin and discolored blemishes that gave him darkish circles under his eyes. He looked haggard. His skin had a "scarf" look that made him feel very tired and very aged, even though he was in his prime of life. His skin hung in loose folds. His chin started to sag and he feared that this would make him look less efficient as an "old" salesman, compared to his younger competitors. He was bothered with frequent outbursts of eczema and skin boils that soon became unsightly blackheads. He recognized that this was Nature's "alert signal" to correct his protein intake or else his body would become vulnerable to more deficiency symptoms.

Skin Firms Up Speedily. Michael A. needed a "complete" protein food. He started eating fish, either boiled, baked, broiled or steamed. (Fried fish was taboo; the frying process coats the fish with a heavy film and renders the protein less digestible. Frying also breaks up the fat globules of the fish protein and its collagen nourishing is less effective.) Michael A. found that by eating fish three to four times weekly, his skin firmed up. Then he made still another protein-boost intake.

Fish Protein + Fruit Vitamins = Cellular Rejuvenation. Michael A. started eating raw fresh fruit with each fish meal. The fruit offered him a powerhouse of Vitamin C, which took the collagen from the protein in the fish and used it to build and rebuild his body cells and tissues. This *combination* of two "miracle foods" stimulated the DNA-RNA manufacture and brought speedy benefits in his skin and his basic health. Life became more youthful.

"Skin Aging Clock" Is Reversed. The use of a "complete" fish protein which is biologically superior to any other type of protein, with Vitamin C helped revitalize his cells. His skin firmed up. Blemishes subsided. He looked youthful. His loose skin became tight.

He had no worries about a "sagging chin" since this skin was as resilient and firm as a youngster, thanks to the cellular rejuvenation made possible by a simple program—*with each fish dish, eat a raw fruit salad for optimum utilization of "complete" protein by Vitamin C.* It is the miracle of body and cellular rejuvenation. It helps reverse the "skin aging clock," thanks to Nature and protein.

HOW TO USE PROTEIN TO REVIVE-RESTORE-REJUVENATE FROM HEAD TO TOE

Just as protein is rejuvenating when eaten, it can also be youthifying when applied externally. To enjoy a radiantly young and attractive skin, no matter what your age, take advantage of these little-known secrets, originating from lore and legend as well as modern folks, which have been culled by eternally young people. The benefit here is that these easy-to-follow, do-it-yourself secrets are prime sources of complete or incomplete protein, together with vitamins and other nutrients, that *combine* in a biological form so as to stimulate and complement the amount of collagen now in your skin. These secrets alert sluggish DNA-RNA components and boost secretion, so that your body becomes alive with tingling life as worn, decayed and dead cells are rejuvenated and replaced. Since your skin is basically protein and almost entirely cell and tissue, it is essential to use protein to keep it looking young from head to toe.

Lemon-Oatmeal Face Lift. About once a week, your face needs a good scrubbing to get rid of all the ground-in dirt and excess oils. It needs replenishment of its used up supply of collagen, sparked by Vitamin C. Here's how to help nourish your skin and help rebuild damaged cells and enjoy a "face lift" feeling that is refreshing.

Mix together the juice of 1 lemon and the white of 1 egg (pure and complete protein). Add dry oatmeal gradually until you have a thin paste. Apply to your face, avoiding areas around the eyes. Let dry (about 10 minutes), then rinse off with clear, warm water. It sends a rich treasure of protein through your pores, activated by Vitamin C from the lemon, and energized by the carbohydrates of the oatmeal. Your cells are "nourished" and plump up your sagging skin, giving you a more youthful appearance. Try this at least once a week.

THAT FRESH-FACED GLOW—WAYS TO GET IT AND KEEP IT

Evelyn T. neglected her personal health as she helped her husband in his business, managed a busy household, took part in many club groups. She began to look prematurely aged. She developed fine lines on her forehead. Her cheeks became sallow. She was aging too fast until she discovered a protein secret that could reverse this reaction and help give her a "fresh-faced glow." Here is her secret:

"Fresh-Faced Glow" Secret: Beat the yolk of an egg (complete protein) until it is light and frothy. Add a half cup of milk (good protein) and a mashed half of a ripe peeled avocado (Vitamin C plus protein and minerals). Beat the mixture with a fork until you have a thin cream or a lotion-like consistency. Apply on squares of cotton. Now keep "washing" your face with this "Fresh-Faced Glow" Formula. Finish off with a light sprinkle of lemon juice.

Benefits: Evelyn T. discovered that this high-protein formula helps replenish the collagen and to stimulate the manufacture of DNA-RNA beneath her skin surface. Once this happened, her face primed up, her color returned, her forehead lines were miraculously erased. Each week Evelyn emerges from this 10 minute "at home" treatment (performed once weekly), looking and feeling like a glowing schoolgirl. It's simple, yet very effective. Protein is the miracle rejuvenator.

PROTEIN FACE MASQUE

Susan W. found beauty shops to be too expensive. Furthermore, the chemicalized lotions scorched her skin, making her look as if she had big blotches. Susan W. had a problem with scaling skin, deep pits and unsightly blemishes, but beauty parlor chemicals made them worse. They burned so deep that her skin cells became more damaged than before. She needed a *natural* method that would rebuild her skin, instead of destroy it. She tried a Protein Face Masque, a secret shared with her by a neighbor who was a grand-mother, yet looked like a schoolgirl.

Protein Face Masque: In a blender, put 1/2 cup cucumber pieces (including the peel) (Vitamin C), 1 egg white (complete protein) and 1 tablespoon dry milk powder (good protein). Whizz until smooth. Now just apply to the face with cotton pads. Let remain as a masque for 30 to 60 minutes. Then remove with lukewarm water. Finish with a light splash of Vitamin C—containing lemon juice.

Benefits: Susan W. found that this super-protein masque sent a shower of collagen-building protein right through her pores, building up her DNA-RNA supply so that her entire face became replenished through new skin cells. Her blemishes soon vanished. Deep pits became smoothed out as if removed. Her skin tone was bright and colorful. This simple protein secret, a *Protein Face Masque,* had given her a "new skin" so that she, too, looked like a glowing schoolgirl!

BERRY-BUTTER BEAUTIFIER

Joan Y. eats a high protein diet plus a balance of other essential nutrients, but her skin still looks pale, has crow's feet in the corners, and the darkish creases that suggest wrinkles. She is able to pep up her skin color and erase the creases and wrinkles by feeding her skin protein from the outside. Here is her secret:

Berry-Butter Beautifier: Mash several strawberries until you have about 2 teaspoons of juice. Blend juice thoroughly with 2 teaspoons of butter. Now smooth onto your face. Work it into your skin for five minutes. Let remain about 10 minutes. Then splash off with soap and warm water. Rinse with cool water to which some lemon juice has been added.

Benefits: Joan Y. has given her tired skin a supply of high-powered Vitamin C from the juice that works with the protein in the butter to enter through the pores, replenish the supply of collagen, so that DNA-RNA can be nourished to build and rebuild skin cells. She finds that this *Berry-Butter Beautifier* works miracles in helping to put a glow to her skin, erase the creases and wrinkles. She uses it once daily and just about glows with radiant youthfulness. (She is 56, but looks a young 36!)

SUPER-PROTEIN FACE LIFT

Barbara D. was troubled with sagging skin. Her neckline dropped giving her an embarrassing "crepe" look. She had unsightly "age spots" and had pale color that made her look fierce! She did not want to submit to plastic surgery because she feared an operation. But she wanted to have all the benefits of a face lift—without surgery. So she created a protein program that worked wonders. It uses just 2 ingredients, but the method in which they are used offers a "million dollars" of rejuvenation, as she explains it. Here's how to have this, *Super-Protein Face Lift* right at home:

Separate one egg (for protein-plus). Add the egg yolk (complete protein) to 2 tablespoons heavy cream (additional protein and vitamins). Stir gently until well mixed. Now beat egg white (pure-complete protein) to a froth. Slowly, stir into the mixture. Next, steam your face (to open the pores and make entrance to the cells more receptive) with a towel wrung out in comfortably hot water. Steam for at least 5 minutes, to open up your facial pores. Remove cloth. Now apply the Super-Protein Face Lift. Rub it all over your face. Let it remain up to 30 minutes. Then wash off with tepid water, followed by cold water. Finish with a lemon juice splash.

Benefit: Barbara D. wisely "opened" her skin pores with a warm towel. This enabled the complete protein of this secret mixture to go directly through the pores, sending a stream of collagen and vitamins to the pores. The DNA-RNA supply could be speedily replenished. Within 30 minutes, her "crepe" look was hardly noticeable. The "age spots" vanished. Her color was softly pink and glowing. She praises this protein replenishment as the closest thing to plastic surgery "but entirely natural!"

LOOK AND FEEL YOUNG ALL OVER—WITH A PROTEIN BATH

For a while, George J. was thought to be the "father" of his wife because she does, indeed, look younger. That was before he decided to do something about his fast-aging appearance. He heard a neighbor say how a famous Mexican health spa performed miraculous rejuvenation by having guests take regular simple protein baths. He tried one himself.

Body Rejuvenation Protein Bath: Fill a tub part of the way with tepid water. Now mix together 1 cup powdered milk (high protein), 1/2 cup ordinary table salt or kelp (a sea salt sold at health stores, a prime source of vitamins, minerals and ocean proteins), and 4 tablespoons peanut oil (good protein) until it is a paste. Then pour into the tub. Mix with your hand to blend in with the water. Now add more water to bathing level. Immerse yourself in this *Body Rejuvenation Protein Bath* for 30 to 45 minutes. Rub yourself all over. Rub this lotion on every part of your body from top to bottom during this 30 to 45 minute soak. Then emerge, let the water run out, splash off under a tepid and then comfortably cold shower. Towel vigorously dry.

Benefits: George J. enjoyed the way the warm water steamed open the millions of pores over his body. Now the protein and vitamins from the ingredients could seep right through, enter his circulatory system, and set about to feed and replenish the needed collagen and then DNA-RNA cellular rebuilding substances. Just 30 minutes and he felt that he was rejuvenated. He looked it, too. After a week of daily baths, his skin glowed with health and vigor. He showed very few creases or tired wrinkles. Now he looked more like his wife's husband, and when someone said he looked like her "kid brother," it was time for George's wife to follow this secret *Body Rejuvenation Protein Bath.*

FOUNTAIN OF YOUTH PROTEIN BATH

Andrew L. rightfully wants to keep up appearances. In a world where youth predominates, he wants to compete with the younger men. He feels "under 30" but he looks "over 50." How can he close this gap? He has discovered a *Fountain of Youth Protein Bath:*

In a cheesecloth or muslin bag, mix 1/2 cup powdered milk (good protein) and 1/2 cup ground oatmeal. (To make ground oatmeal, just put ordinary breakfast oatmeal into a blender and blend until it feels powder-smooth to your fingers. One cup of oatmeal will give about two-thirds cup ground oatmeal.)

Fill a tub with very hot water. Drop the bag into the tub. When the water is cool enough to soak, then squeeze this bag and swirl back and forth to further disperse the ingredients. Then soak yourself up to 45 minutes. Let the water swirl all over. Rub yourself with the protein-ized water. You'll soon feel exhilarated. Let the water run out. Finish off with a tepid and then comfortably cool shower. Rub vigorously with a turkish towel.

Benefits: The warm water opened up the millions of pores over Andrew L.'s body. The protein from the milk and oatmeal now gained easy entrance to his skin cells. The collagen from the protein stimulated a healthy production of DNA-RNA so that his billions of skin cells could be repaired. He emerged revitalized, and always said it was like a 45 minute soak in the Fountain of Youth. Now he felt "under 30" and looked "under 30," too.

Thanks to the miracle collagen-producing power of protein, you can revive-rejuvenate-replenish your skin from head to toe with this simple nutrient. Use protein inside—and outside—and be rewarded with a "forever young" skin.

IN REVIEW

1. Collagen, a protein substance that is considered a miracle nutrient that can replenish "aging" skin and plump up sagging wrinkles.

2. Plant protein, such as in Brewer's yeast, helps give the skin cells the needed working materials for making DNA-RNA. These are the "magic" substances that can keep body cells perpetually rejuvenated.

3. A doctor helped reverse the aging process in his patients by giving them a balanced high-protein program with emphasis upon Brewer's yeast. Wrinkles vanished. Skin smoothed out. Mental health responded, too.

4. To feed your skin cells needed collagen, try any or all of the listed 10 miracle protein skin food combinations.

5. Gladys G. found that a protein feeding program gave her a natural "face lift," within three weeks.

6. Fish is an excellent source of "complete" or "superior" protein, and a treasure of skin-rejuvenating collagen.

7. Michael A. *combined* fish protein with fruit vitamins for a unique cellular rejuvenation that reversed the "skin aging clock."

8. A lemon-oatmeal home treatment helps give you a "face lift" reward.

9. Evelyn T. uses a *Fresh-Faced Glow Formula* (high protein) to look like a schoolgirl.

10. Susan W., brightens up with a *Protein Face Masque.* Joan Y. uses a *Berry-Butter Beautifier* to erase wrinkles. Barbara D. uses a *Super-Protein Face Lift* to do away with sagging skin "crepe" look. George J. and Andrew L. use different "Fountain of Youth" protein baths to emerge looking like youngsters, thanks to the secret of collagen and DNA-RNA, from protein!

3

HOW TO FEED PROTEIN TO YOUR SCALP FOR HEALTHY HAIR

Your hair is protein!

A single strand is composed of 18 of the known 20 amino acids found in the body. Hair is made of the same kind of tissue as the outer layer of your skin, or epidermis. Your scalp is skin and it consists of nerves, sweat glands, blood vessels, sensory apparatuses for heat and cold, sensory cells at the ends of the nerve fibers, sebaceous glands, muscles and millions of cells and tissues.

Feed your scalp complete protein and you offer nourishment to your hair which may be considered a complete protein, too.

HOW PROTEIN IS AN ALL-NATURAL "HAIR FOOD"

When you protein-ize your body, and when you use protein applied externally to your scalp, you send amino acids to the epidermis (outer skin) and to the dermis (inner or "true" skin). These amino acids then "feed" your hair by entering the tiny indentations in your scalp (about a quarter million of these), and nourish the follicles at the base of which is cellular tissue containing blood vessels.

The amino acids then seep into the blood vessels which use these nutrients as actual food for your hair. Since hair is basically protein, it is a wise plan to feed it protein for youthful health.

Protein Nourishes Complete Hair. Your protein hair is constructed of three layers. Each layer needs protein for nourishment. The *cuticle* is a shingle-like formation of keratinized protein. The *cortex* uses protein to maintain the healthful amount of moisture. The *medulla* needs protein for strength and vigor. An adequately protein-ized body means that all components benefit. These proteins are then sent to your scalp to nourish the tissues and then nourish the

43

hair, itself. Give your scalp and hair adequate protein and you should enjoy a healthy mane that is a crowning glory and a symbol of healthy youth—at any age.

COMPLETE PROTEINS ARE THE SECRETS FOR HAIR HEALTH

Complete proteins are a vital constituent of every cell of your scalp, making hair growth possible and furnishing the building blocks needed for nourishing the hair itself. Complete proteins are those which have all the essential amino acids. (Incomplete proteins are those which lack one or more essential amino acids.) To give your scalp and hair cells a *complete* balance of *all* amino acids, you need complete protein foods. Since your hair contains almost *all* amino acids, it is essential to feed them *all* amino acids. A deficiency of one amino acid (as in an incomplete protein) may create a shortage that is contrary to hair health.

Official U.S. Government Recommendations for Complete Protein Foods. Dr. Ruth M. Leverton, in *Food*, tells us:[1]

"The presence in a protein of all the essential amino acids in significant amounts and in proportions fairly similar to those found in body proteins classifies it as a complete protein—meaning that it could supply completely the needs of the body for these amino acids."

Balance Needed. "The proportions in which the essential amino acids are required are as important as the amounts. Apparently, the body wants these amino acids to be available from food in about the same proportions each time for use in maintenance, repair, and even growth."

List of Foods. "*Meat, fish, poultry, eggs, milk, cheese, and a few special legumes (chickpeas, soybeans) contain complete proteins.*"

Gelatin Is Incomplete. "Gelatin is the only food from an animal source that does not meet the specifications."

From the preceding list, we see that the *eight* foods are recommended by the U.S. Government as being *complete* proteins. Since hair and scalp are almost complete protein, your "hair foods" *daily* should consist of one or more of these *eight* government recommended foods.

Complete Protein Is Essential. A deficiency can mean scalp-hair deprivation. Dr. Ruth M. Leverton adds, "When incomplete proteins are metabolized, they supply the body with enough of *some* of the essential amino acids but inadequate amounts of others. *The amino*

[1] *Food, The Yearbook of Agriculture,* Ruth M. Leverton, United States Department of Agriculture, Washington, D.C., 1959, pages 66-67.

acids that are supplied are not used unless the other essential amino acids are present from other food sources. Instead, they are oxidized, and the nitrogen portion is excreted—the amino acids *cannot* be stored for use at a later time when other acids become available." So complete protein is needed daily in order to nourish your billions of skin cells. Your scalp contains those cells which need *complete protein nourishment* to feed the hair which, itself, is almost complete protein.

A 10-STEP HAIR FEEDING PROGRAM

Dorothy C. was always on the go. An active clubwoman, she had to travel, "eat on the run," manage a household, and go from one chore to the other. This constant tension drained out many valuable nutrients. She showed a protein deficiency in dull, lifeless, stringy and thinning hair. She had bouts with dandruff-flaking dry hair, to greasy oily hair. Dorothy C. thought she could have instant hair beauty at the local parlor. But it only worsened her condition. Bleaches, tints and toners only broke down her hair structure and destroyed protein content in her scalp. She looked haggard. She had to do something.

A concerned friend who always had thick, healthy hair, took her aside, urged her to cut down on her hectic activities. Then she outlined a simple but effective protein-nourishing 10-step hair feeding program. Dorothy C. tried it for four weeks. She bounced back with amazing strength and vigor. Her hair actually looked thicker. It glowed. She had no problems of dandruff or oil. Her scalp felt tingling and alive. She is grateful to the friend and to protein for this hair feeding program. Here it is:

1. *Complete Protein Foods Several Times Daily.* To nourish the scalp tissues and cells and give them amino acids for hair feeding, eat several complete protein foods daily. Examples include a soft boiled egg for breakfast, cottage cheese for lunch, lean meat or fish for dinner. Frequently, eat chickpeas or soybeans. Vary the list of complete protein foods for better taste and better balance.

2. *Eat Vitamin Foods for Better Protein Assimilation.* Your scalp tissues cannot send "bulk" protein into your hair shaft or into the hair itself. That protein must be metabolized and assimilated by vitamins and enzymes found in raw fruits and vegetables and their juices. Daily, eat large, raw salads and drink juices. This sends a shower of Vitamin C and enzymes into your bloodstream, directly to your scalp tissues where the protein is now broken down into amino acids that become food for your hair.

3. *Seed Protein Helps Provide Natural Moisture.* Beneath your scalp surfaces are millions of tiny cells and tissues requiring moisture. The essential fatty acids found in seeds are liquids needed for your thirsty scalp cells and tissues. Give them this liquid and they help protect against excessive dryness or dandruff. Daily, eat raw seeds (sunflower, sesame, pumpkin are just a few that are good moisturizers), nuts and wheat germ. Use these seed oils in a raw vegetable salad, too. You'll be giving your scalp tissues the needed protein and the moisturizers required for better hair health.

4. *Eliminate Sugar and Bleached Flour Products.* Refined sugar products and bleached four foods cause an unhealthy and erratic metabolism that can deplete the protein supply going to your scalp tissues. Scaly scalp tissues indicate that these bleached carbohydrate foods are displacing protein. Eliminate them totally from your food program for better protein nourishment of your body cells and your hair shaft.

5. *Raw Foods Are Prime Sources of Nutrients.* Raw fruits and vegetables contain vitamins and enzymes that are needed to *activate* protein and *metabolize* it into usable amino acids. Cooked foods are very low in these essential protein boosters, and your scalp tissues and hair are correspondingly "cheated" of needed amino acids. A rule of thumb is to eat raw what can be eaten raw. Cook *only* if the food *must* be cooked. It is the natural way to supply nutrients that boost protein metabolism and encourage better scalp-hair health.

6. *Whole Grain Foods Are Important Protein Sources.* Whole grain breads and cereals are prime sources of amino acids that are different from animal foods. You need them for a *balance* of different amino acids to give your scalp and hair a *variety* of nutrition. Eat them daily.

7. *Natural Foods Are Best for Tissues.* Your scalp tissues need natural and "living" food to feed your hair "living" protein. Avoid foods that have chemical preservatives and sprays. These are "dead" and their protein content is destroyed. Avoid artificial sweeteners. Replace coffee, tea, cocoa, chocolate and carbonated drinks with coffee substitutes, herbal tea, carob drinks and fruit juices. Health stores have these as well as many supermarkets, too. They are prime sources of "alive" protein for "alive" hair nourishment.

8. *Boost Scalp Nourishment with Raw Salads.* To give super-metabolism of protein, *begin* each meal with a raw fruit or vegetable salad, or a raw juice. The *benefit* here is that this sends a supply of vitamins and enzymes into your digestive system so that protein food later eaten is better assimilated. Amino acids are more abundant for your scalp and body tissues with this easy method.

9. *Plant Proteins Are Important.* Daily, eat fresh, green leafy vegetables as well as whole grains and seeds for unique protein. Plant proteins have a special amino acid pattern unlike animal proteins, and your scalp will benefit from this combination of vitamins-enzymes-amino acids in a plant food. Eat plant foods daily.

10. *Dairy Products Are Good Protein Sources.* Dairy foods, such as milk, buttermilk, yogurt, natural cheeses and cottage cheeses are prime sources of

special types of protein that help stimulate scalp and hair health. Dairy is the "almost perfect" protein food and should be eaten daily for better body and scalp tissue nourishment.

This 10-step hair feeding program is the *foundation* for giving your scalp tissues needed amino acids for hair nourishment. Build it into your daily living program and you'll reap the rewards with a feeling of better health from the top of your thick, healthy hair to the bottom of your smooth skinned toes!

HOW A SIMPLE FRUIT VITAMIN ALERTED SLUGGISH SCALP PROTEIN

Paul E. had an embarrassing problem. He had "tight scalp" that was so dry and itchy, just a slight touch would produce unsightly "snowflakes" of dandruff. No matter what he tried, his scalp still was dry. He felt discouraged until he learned that proteins needed to be stimulated by Vitamin C to create collagen, a health skin tissue that could then send amino acids into the hair shaft. But *without* Vitamin C, protein could remain inert and unable to be transformed into amino acids.

Simple Fruit Vitamin Rinse. He squeezed the juice of a half lemon into a basin of water. After a shampoo, he would then rinse his scalp and hair in this vitaminized water. *Benefits:* The Vitamin C in the lemon juice helped to restore the "acid mantle" of the scalp skin. In this environment, protein became metabolized into amino acids that nourished the cells and tissues, corrected dryness and promoted better health of the scalp and hair. Now, Paul E. uses this easy *Fruit Vitamin Rinse* to activate protein and create a healthy scalp for better hair.

Protein Needs an Acid Balance. Your scalp tissues and cells can receive protein but may not metabolize them without the necessary acid balance. Here is how it works. Human hair and scalp tissues normally have a pH (acid-base) that varies from about 4.5 to 5.5. At a pH of 4.5, the scalp tissues are able to completely metabolize protein into amino acids which feed the hair. At this pH of 4.5, the amino acid nourished hair has its greatest elasticity, strength and luster. Lemon juice or any acid juice from a fruit can create this pH of about 4.5 to stimulate protein metabolism. It is the "acid balance" that creates healthy scalp tissues and better hair health. It is the secret for better cell-tissue rejuvenation on your scalp.

5 WAYS TO FEED PROTEIN TO YOUR SCALP TISSUES AND HAIR

Five common scalp-hair problems may respond when protein is given for nourishment.

1. *Dry Hair Mayonnaise Healer.* Shampoo and towel dry your hair. Now apply about one tablespoon of mayonnaise and leave it on your hair for one hour. Then shampoo lightly and rinse in lemon-water. Its protein will seep through your pores and nourish the thirsty cells and tissues. Its amino acids will nourish your hair. It helps correct dryness of moisture-starved scalp and hair.

2. *Skim Milk Protein Wave Set.* To add protein and body to your hair, use skim milk as a wave set. Apply to your hair. Do not rinse out but allow to dry. The protein will firm up your hair and your curls will remain days longer.

3. *Protein for Oily Hair.* Dissolve one tablespoon salt in some skim milk and rub into your scalp. Do not rinse out. Let it dry. It helps send protein into your scalp but the salt acts as a "sponge" for excessive oil and this helps create a better balance for better scalp health.

4. *Protein for Thick Hair.* If hair is thin, or if it is a "problem," you can feed it needed protein. Mix dried skim milk powder with water to make a *paste.* Apply to your hair as a thick pack. Let it remain up to 30 minutes. Then wash out. Milk protein goes into your scalp pores, nourishes your tissues which can then use it for feeding your hair. It helps ease "problem hair" and helps provide tensile thickness.

5. *Wheat Protein for Frizzy Hair.* For very "frizzy" or dry hair, try wheat protein. It has a special combination of amino acids that appear to nourish and replenish starved scalp tissues. Just apply two tablespoons wheat germ oil to your hair. Massage into your scalp. Comb through. Brush vigorously. Repeat several times a week to moisturize the "thirsty" scalp tissues and cells.

Let protein solve your scalp problems and you'll enjoy the look and feel of hair-growing tissues.

HOW TO USE PROTEIN AS A HEALTHY SCALP-FEEDING SHAMPOO

Carol F. looked a fright. She said she had one of those heads of hair that nothing could be done for. She has submitted to chemicalized bleaches, tints and dyes until her hair structure was completely broken down. Worse, she started using so many chemical hair sprays, that she looked sprayed and lacquered like a piece of furniture! She had to do something to feed her starved scalp tissues the needed protein that harsh detergent shampoos, alkali soaps and chemicalized beauty preparations had just stripped away. She was given two secret or little known protein shampoo recipes that worked wonders. Here is what Carol F. did. Twice a week she would shampoo with these protein feeders. She would alternate them. This sent a *complete* protein to her starved scalp and tissues and her hair regained natural lustre. Now she uses them regularly and avoids any chemicals on her head and hair. She has an enviably thick head of healthy hair. Here are the two secret or little-known shampoo recipes you can try at home.

I. Complete Protein Shampoo

1 egg
1 cup castile shampoo

Put the egg into a blender and whirr until well beaten. Now remove cap and slowly pour shampoo in a steady stream. Use a rubber spatula if necessary to keep ingredients around processing blades. Shampoo as usual.

Benefits: The *complete* protein of the egg offers all essential amino acids to your scalp, replenishing the starved tissues and cells. In particular, the *complete* amino acids nourish the medulla portion of the hair shaft which is the source of protein tissue food.

II. Super Protein Shampoo

1 egg
1 cup castile shampoo
1 package unflavored gelatine powder

Put the egg into a blender and whirr until well beaten. Now remove cap and slowly pour shampoo in a steady stream. Add gelatine and continue to process until mixed well. Use rubber spatula if necessary to keep ingredients around processing blades. Shampoo hair as usual.

Benefits: The *complete* protein of the egg and the nearly complete protein of the gelatin *complement* one another to create *protein-plus nourishment* of your scalp tissues. This unique combination adds body to your hair and helps condition it . . . naturally. The amino acids also nourish the photofibrils of your hair. These are like splinters that need the amino acids from the *Super Protein Shampoo* to tighten the hair structure back to normal. It is believed that this nourishment offers better tensile strength and better thickness.

HOW WALTER J. USED A *SECRET SCALP TONIC* FOR ENDING DANDRUFF

Walter J. had a lifelong problem with dandruff. Not only did his shoulders show the "flakes" but he had an agonizing scalp itch. He also had tightness of his scalp which made him feel all tensed up and nervous. When he scratched his scalp, his fingernails were clogged with dead scales. This was a *clue* to his problem. That is dead scales suggest dead scalp tissues! Why did they die? *Lack of protein nourishment!*

Walter J. ate a balanced diet that was nourishing with protein. But the protein needed an *activation* from the *outside* by Vitamin C. (This same inter-action is made possible during the digestive system when enzymes take Vitamin C from fruits to metabolize the protein from eaten foods. It can also occur externally.) Walter prepared an amazingly simple but highly effective *protein booster* that he called a *Secret Scalp Tonic:*

> Squeeze the juice of one lemon into a bottle of castile or organic shampoo. Mix thoroughly. Then shampoo as usual. Finish by using a rinse of the juice of half a fresh lemon in a glass of warm water.

Benefits: The Vitamin C in the lemon juice acts upon the scalp to cleanse and metabolize the protein so that it can be transformed into *living amino acids* to nourish the hair follicles and feed the hair.

Results: Walter J. experienced a tingling clean and deliciously refreshed scalp. His hair fairly glowed from this *Secret Scalp Tonic.* Gone was his itching scalp. Gone were the ugly dandruff scales or dead scalp tissues. Instead, he had a living scalp that helped produce living hair, thanks to the interaction of Vitamin C and protein.

TIP: For top-notch vitamin metabolization of protein on the scalp, just squeeze the juice of a lemon directly onto your hair. Use a protein shampoo and then rinse in water to which a little lemon juice has been added. It's the natural way to wake up your scalp and metabolize protein for your hair!

HOW PEANUT OIL CAN GIVE PROTEIN BODY TO YOUR HAIR AND SCALP

Peanut oil is a prime source of protein. Two tablespoons of cold-pressed oil may have as high as 20 grams of protein. This is the complete amino acid balance your scalp and hair thrive upon. Here is how you can actually feed this high-quality protein directly to the roots of your hair. The amino acids go directly to the "thirsty" cells and tissues on your scalp.

Peanut Protein Plan: Warm two tablespoons of peanut oil over hot water until comfortable to your fingertip touch. Gently massage this into your scalp, using the bony part of your fingers. Do not use your fingernails as this will scratch the membrane of your scalp tissues. Rub thoroughly *into* your scalp. The warmth will *open* your pores so this treasure of amino acids can enter to be "eaten" by your tissues and cells. Now wring out a towel in hot water. Wrap around your head as a turban.

After it cools, repeat this *Peanut Protein Plan* several more times to insure that your tissues are saturated and their "thirst" quenched. Then use a Protein Shampoo. To your final rinse water, add some lemon juice. Do this several times a week.

Benefits: When your scalp pores open, the amino acids from the peanut oil go directly to your millions of cells and tissues which "drink" this nutrition, later to transport it to your hair through the bloodstream. It helps keep damaged hair from breaking. It offers nutrition to your scalp, unlike any other external protein-ization method. It will be welcomed by your scalp. It will reward you with better health of your hair.

SECRETS OF BETTER SCALP-HAIR HEALTH

Give protein a helping hand with overall better body and hygiene care. Here are some ways to help grow healthier hair.

Improved Food Program. A natural diet helps grow more natural health. A natural diet that thrives in protein will offer thriving health to your scalp-nourished hair. Living food means living hair. Avoid processed, devitalized and *refined* carbohydrates. Avoid refined sugar in any form. These carbohydrates cause erratic gland behavior and protein is displaced. Eat wholesome foods.

Regular Brushing. The old-fashioned 100 nightly strokes helps "wake up" the protein so that scalp glands can better metabolize them into amino acids. Use a natural bristle brush to boost scalp circulation and better distribute amino acids to your millions of cells and tissues. Use even movements during brushing. Start at the nape of your neck, slowly brush all the way around your head until you return to the starting point. Brush *outward* from your hairline. Cover every portion of your scalp.

Keep Relaxed. Tension tightens up your scalp. Protein becomes "locked" and cannot become suitably metabolized. Relax yourself and you'll find a better distribution of amino acids adds up to better scalp and body health, too.

Self-Massage. Stimulate your scalp glands so the roots can "drink" the amino acids. Often, a sluggish scalp means "starved" roots. Wake up your scalp by placing palms flat against it. Close your fingers around the hair at its lowest point. Gently pull several times before moving on to another area. This alerts your scalp, wakes it up, and helps distribute amino acids to as many "thirsty" cells as possible.

Natural Hair Cosmetics. Avoid excessive use of chemicals since these usually contain alcoholic solvents that destroy the vital proteins and the sulfur bonds, forcing the molecular strands to become separated. (A protein treatment will act as an adhesive to bring together these divided strands.) Avoid any detergent-type shampoos, alcohol-based lotions or chemical sprays. Use above-described *natural* hair foods to help nourish your scalp tissues. Chemicals such as dyes or tints only destroy the scalp tissues and burn up protein. Natural *foods* for the hair will nourish your tissues and give you better health of your scalp and hair.

Easy Dandruff Remover. Mix 1/2 cup white vinegar with 1/2 cup water. Using absorbent cotton pads, dab this solution right onto your scalp *before* shampooing. It is able to remove dead scales and tissues, and after shampooing, enables the protein (from your shampoo) to gain access to your thirsty scalp tissues much easier. Do this several times a week and always before shampooing until dandruff has gone.

Natural Protein Conditioner. Beat 1 egg yolk. Add 1/2 cup yogurt and beat until foamy. Gently comb this *Natural Protein Conditioner* after you have shampooed and towel-dried your hair. Let it soak in for 30 minutes during which time the protein is absorbed by your thirsty scalp tissues and amino acids are sent to your hair. Then rinse out thoroughly in warm water. A final rinse has a bit of lemon juice added. You will find that the protein nourishment has given extra body to your hair—the natural way!

Your scalp (and body skin) is 97 percent protein. Nature gives you a protective barrier of friendly acid containing oils and special moisturizing ingredients. Maintain this acid-balanced barrier. Avoid alkaline soaps and creams which remove this protective barrier. By using the protein programs outlined, you'll help keep your scalp soft, smooth, glowingly alive and slightly acid in reaction. Protein will help keep your scalp tissues in a delicate moisture balance and protect against the tough-flaky dryness of premature aging. Your hair and scalp are made of protein. They feed upon protein. Give your body a balanced food program that is high in protein and you'll be rewarded with a healthy scalp and healthy hair growth—thanks to Nature.

SUMMARY

1. Protein is an all-natural "hair food" that nourishes the complete hair. Complete proteins, as recommended by the U.S. Government, offer all the amino acids needed by your scalp tissues and hair.

2. Dorothy C. followed a tasty 10-step hair feeding program to invigorate her cells and tissues. She was rewarded with a youthful appearance and healthy hair.

3. A simple vitamin found in an everyday food available for a few pennies helped alert protein so that Paul E. corrected his problems of "tight scalp" and dry, itchy "snowflake" dandruff. This vitamin-protein combination restored the acid balance, the key to youthful hair.

4. Discover 5 ways to feed protein to your scalp tissues and hair, for different common problems.

5. Carol F. uses two protein shampoos as "food" for her scalp. She has lovely hair as her reward.

6. Walter J. ended dandruff with his easy *Secret Scalp Tonic.*

7. Peanut oil gives protein body to your hair and scalp. Feed your tissues easily with the described program. The *Peanut Protein Plan* nourishes your tissues and cells within moments.

8. Check your hair-scalp problem and follow the secret ways to protein-ize your tissues for better health.

4

REBUILDING YOUR BRAIN CELLS WITH NATURAL PROTEIN

You are as young as your brain cells, all 16 billion of them! Feed them natural protein and they will keep renewing themselves, and keep giving you an alert and youthful mind. When natural protein nourishes these brain cells, they are able to reproduce properly and become part of the body network that enables thoughts to work with youthful speed from head to toe.

How Protein Acts As a Mind Energizer. Metabolized protein sends a supply of amino acids directly to the neurons or nerve cells connected to the brain. These nerve cells use amino acids as nourishment to energize them to transmit "messages" for the control of your body, for your behavior. The amino acids send these "messages" from neuron to neuron along the passageway. Amino acids become "transmitters" and enable your brain to dictate orders to be carried out by your body.

Youthful Mentality. Amino acids perform in the way that certain fluids conduct electricity in batteries. They nourish the nerve cells and tissues of the brain, and alert them to activate the thinking segments so that the body can then fulfill the orders of the nourished mind. A deficiency of protein means that the nerve cells become "starved" and they cannot function as adequately as if properly energized. It is especially important for the brain to be protein-nourished in the middle years. The reason is that the brain cells die more profusely after the age of 30, suggesting that Nature wants stronger protein rebuilding. (The average brain shrinks in weight from 3.03 pounds at age 30 to below 2.00 pounds at age 50. It can keep losing brain cells, and reduce the thinking powers, if protein nourishment falls below par.) The key to youthful mentality lies in adequate protein for the brain, and the entire body, especially in the third decade of life. It offers hope for a "think young" ability for many, many more decades thereafter.

HOW A HIGH PROTEIN BREAKFAST DOUBLED THINKING POWER

Jeff H. was an overworked accountant. Obligations kept multiplying as did the figures on his balance sheets. He could not refuse these assignments because so many people depended upon him. Jeff H. was tops in his field, so when errors began creeping on his forms, when costly mistakes threatened to ruin his career, he knew that something serious was wrong.

Poor Breakfast Reacted in Shrinking Brain Cells. Because Jeff H. was always in a hurry, he had to gulp down breakfast. It consisted of some sugary sweet rolls, a pastry, coffee with sugar. By mid-morning, he felt "in a slump" as long columns of figures became blurred. His brain refused to work. He garbled up numbers. His thinking was fuzzy. Even his words were slurred. He felt as if his brain actually stopped working! The reason appeared that his high starch and *no* protein breakfast had contributed to hypoglycemia or "low blood sugar" wherein the brain cells become starved and cause erratic emotional behavior patterns. Jeff H. had to correct his breakfast and general protein-izing of his brain and body.

Protein Feeds Brain Cells. Jeff H. was to emphasize *protein* for breakfast. Each morning he would select *one* of these "complete" protein foods: meat, fish, eggs, cheese, peas, soybeans or many nuts. With this protein food, he would have a fresh citrus fruit so that its Vitamin C would be able to metabolize the protein into amino acids for use by his billions of brain cells. He would also have a carbohydrate food such as toast or a whole grain cereal (for additional protein, too). Coffee was avoided because the caffeine caused ambivalent responses to his thinking and emotional components. Jeff H. found that this healthfully high-protein breakfast gave him a feeling of refreshed mental power. He continued eating protein for lunch and for dinner. He enjoyed a variety of protein foods throughout the day, in harmony with vitamins-mineral-enzyme-carbohydrate foods. The emphasis was on protein.

Jeff H. learned that a high protein breakfast just about "doubled" his thinking powers. He saw long columns with a steady eye and had a steady hand. He was alert, youthful and could think young. Life had more joyful meaning, thanks to a protein-nourished brain!

Feeds Essential "Thinking" Centers of Brain. When protein is metabolized into amino acids, these nutrients speedily feed the three essential "thinking" centers of the brain:

1. *Cerebrum.* This frontal part needs protein for nourishment of its mass of grey matter. Protein enables sensation of all kinds to be received here and then distributed to reflex pathways or to the cerebral cortex and consciousness. Protein feeds the billions of cells in the "grey matter" to help alert the pituitary gland to issue hormones that help keep you "thinking young."

2. *Cerebellum.* Protein feeds this "hindbrain" that lies in the lower back part of the skull. Protein replenishes its billions of cells which are then able to regulate equilibrium and help coordinate muscular movements. A deficiency of protein may cause a staggering gait, trembling muscles, slurred speech and emotional upheaval.

3. *Medulla Oblongata.* Here the brain requires a constant supply of protein to nourish its nucleic acids, the materials which are used in the form of DNA-RNA to build and rebuild new cells. Well nourished cells in the medulla oblongata nourish the extension and upper terminal of the spinal cord. The protein is used by twelve cranial nerves originating near this region. A well-nourished medulla will help create better communication between the spinal cord and the brain. When the billions of cells are replenished through adequate daily protein, they are able to offer better breathing, a better heartbeat, better swallowing. The nourished medulla oblongata rewards the body with a smoothly coordinated mind. It helps you think young!

Brain Is Largely Protein. The brain, itself, is largely protein and consists of nerve cells and tissues that need a large and constant supply of amino acids. A well-proteinized brain is able to organize the body functions so that you can walk, talk, think, sleep, and do just about everything, with youthful vigor. The *secret* of an eternally youthful mind is a protein-ized brain!

HOW PROTEIN SNACKS GIVE YOU "DAY LONG" MENTAL VIGOR

Lillian I. has her hands (and mind) full from morning to late evening. She has to get her brood off to school in the early morning. Then she has to prepare her husband's papers for his study on he commuter train. As soon as they're off, she hurries around the house to put things in order, and then she has the daily marketing to do, along with meal planning and preparation. She also does research work in the library for her husband to help make his obligations less strenuous. Evenings, Lillian I. is constantly on the go, feeding her family and herself, and discussing the research with her husband. She is more than a wife, mother and housekeeper. She is an "associate" for her husband's business. Lillian I. has a secret for her amazing mind-body energy.

Secret of Protein Snacks. She eats a well-balanced diet. But she needs something "extra" to keep her mind going. She discovered a secret. Snack on high-protein foods throughout the day. Lillian I. uses these "brain foods" to keep her going: Cheese squares on whole grain bread squares. A turkey drumstick. Cold chicken slices. A hard-boiled egg with a fresh fruit. Assorted seeds and nuts for munching. A cup of baked soybeans. Brown rice pudding with sun-dried raisins.

The Super-Protein Food with Dynamic Brain Power. Lillian I. is cholesterol conscious and is seeking a super-protein food that has *no* cholesterol. She has discovered one such miracle food. *It is egg white!* She scrambles several *egg whites* in corn oil, sprinkles them with some Brewer's yeast flakes or wheat germ, and will eat this with a fresh fruit. Lillian I. finds that this *Super-Protein Food* gives her "dynamic" brain power almost immediately. An extra bonus is that egg whites have *no* cholesterol and may be eaten regularly for brain boosting.

Secret of Brain Stimulation Power: Egg white is a *complete* protein. It contains *all* essential amino acids and even those that are non-essential. It also contains vitamins and minerals. When taken with a fresh fruit, the Vitamin C acts upon the protein to transform it speedily into amino acids which are then used by the body cells; but the Vitamin C also sends these needed amino acids to the brain for quick replenishment. Since daily stresses and tensions deplete the brain cells more than in any other body part, it is necessary for ample protein to be always available. This *Super-Protein Food* nourishes the brain so that its cells are quickly rebuilt and the thinking processes are kept youthful and alert.

Lillian I. is able to have that youthful "get up and go" because she uses this *Super-Protein Food* and also has frequent protein snacks throughout the day. Her well-nourished brain cells have given her this youthful zest.

THE PROTEIN PUNCH THAT GIVES YOU "BRAIN POWER"

Robert K. is a factory supervisor and must keep on his "mental toes" all the time. He found his thinking fuzzy. He made costly errors. He was forgetful. Someone said he was getting senile. His thoughts were not too collected. He frequently started on a task, then forgot what he was doing! Robert K. was a heavy coffee drinker. This gave him an "up and down" stimulus. But a protein deficiency was the root cause of his problem. So instead of the "coffee break" he was told to take a "protein break" in the form of a special brain tonic:

Protein Punch for Brain Power. Mix one tablespoon of Brewer's yeast (about 20 grams of complete protein) in a glass of fresh fruit or vegetable juice (for vitamins-minerals-enzymes). Stir vigorously. Drink in mid-morning. Then drink another glass in mid-afternoon. Drink a third glass of the *Protein Punch For Brain Power* in early evening. Together with a more healthful food program of vitamin foods, mineral foods, carbohydrate foods (he ate natural carbohydrate foods such as fruits, vegetables, whole grains and avoided refined starches), and protein, he was able to bounce back with youthful vigor. Now Robert K. is able to think youthfully. His thinking is as alert as a youngster. He enjoys a sharp memory. In fact, his work was so excellent, he was soon given a promotion. Protein made a new man "and a new mind" out of him!

Benefits: The complete protein of the Brewer's yeast is speedily transformed into amino acids by the fruit or vegetable juice enzymes and vitamins. The amino acids are sent via the bloodstream for quick absorption by the brain cells. Here, the amino acids energize the DNA-RNA nucleic acids to create new brain cells and to correctly repair those that have been damaged. The proteins in this *Protein Punch For Brain Power* become the building blocks and components of the billions of brain and body cells. They help rejuvenate the body, inside and outside! Just two ingredients, but a powerhouse of brain regeneration!

HOW GRAIN FOOD PROTEINS HELP "WASH" YOUR BRAIN CELLS

Living long is the reward for following the laws of Nature. But it is more essential to live youthfully in a healthy body . . . that has a youthfully healthy brain!

Brain aging is reportedly traced to a loss of cellular reserves as well as connected to the problem of "littered brain cells." This is a little-known cause of so-called "brain aging." Cellular aging is the consequence of a biological process whereby an oxidation process breaks down cells, tissues and molecules and creates what is known as a littering of "free radicals" or roots. These so-called "free radicals" are waste substances that result from a chain reaction of molecular disintegration.

Cell Littering Causes Aging. These waste substances, or cell litters cause aging by reacting with DNA-RNA, the nucleic acids, which repair body cells and tissues. These cell litters tend to clutter up the billions of body cells, especially in the brain which appears to be affected more than most other body parts. Because these "free radical" wastes interfere with the normal DNA-RNA replication of new cells, they actually hasten on aging of the brain, and body. A key to brain perpetuation would be in helping to "wash out" these "radicals" or waste substances so that cellular rejuvenation can continue on without interference.

Why Grain Proteins Have This Brain Cell Washing Effect. Grain proteins are joined by another unique substance in the plant—Vitamin E! In this Nature-created harmony, protein takes the Vitamin E and uses it to control cell litter via the oxygen circuit. The combination of protein and Vitamin E helps prevent oxygen from combining with fatty acids to form peroxides or "radicals." The protein uses Vitamin E to help get much more mileage out of the oxygen that your blood gathers from your lungs. By helping to "wash" the wastes out of your system, your brain (and body) cells can now use the amino acids for replication and promote a more youthful brain with repaired and functioning cells.

Secret of Grain Protein Brain Cleansing. Whole grain foods are prime sources of Vitamin E, while other protein foods have little or almost no Vitamin E. But Nature has put *both* protein and Vitamin E in grain foods so they can exert their antioxidant therapy, whereby the free floating "radicals" or "litter" from decaying and dead cells can now be washed out. Because Vitamin E has this antioxidant factor, it makes it unique. Almost no other nutrient can do this "washing." Grain Protein takes this Vitamin E which has the antioxidant factor to help wash out the decaying litter so that the bloodstream is sparkling clear. The brain cells are now able to replicate *without* interference by the debris of the "radicals."

A SIMPLE BUT REJUVENATING "BRAIN CLEANSING PROGRAM"

Elizabeth M. finds that she is able to "think young longer" by following this easy "Brain Cleansing Program" right at home, using protein and whole grain foods:

1. All fats are from *plant oils* only! For salad dressings or cooking, she uses any type of vegetable, fruit or grain oil. She uses sunflower seed oil, peanut oil, corn oil, safflower seed oil, cottonseed oil or mixed plant oils. This gives her body a treasure of Vitamin E and protein to be used for the washing out of cellular debris.

2. Daily, she will use wheat germ as part of a whole grain breakfast cereal made with fresh fruit slices. She uses wheat germ in cooking, too, as a filler for meat loaves, in chopped salads and as part of a salad dressing with wheat germ oil. This gives her a good supply of ready-to-use Vitamin E that acts as a "scrub brush" by protein to cleanse away debris.

3. Cooked foods include liver that is sprinkled with lemon juice and wheat germ, and they are enjoyed with a large raw salad containing a dressing of vegetable oil and some fruit juice. Now she is able to feed herself a combination of Vitamin E and protein that gives her healthy vigor. More important, they work to build and replicate her brain (and body) cells. She can work and think like a youngster.

4. For in between meals, she enjoys peanuts, walnuts, assorted seeds and nuts. She also emphasizes that all grain foods should be whole grain, unbleached and non-processed. The secret here is that Vitamin E is perishable and processing can deplete and destroy its cell-washing power. For top-notch Vitamin E "cleansing" power, your foods should be as wholesome and non-processed as possible.

5. Elizabeth M. has a *Morning Brain Booster*. In a glass of fresh orange juice, she adds one tablespoon of peanut oil and a sprinkle of lemon juice for a piquant taste. She drinks this every morning. *Benefits:* In the morning, the brain cells are in greatest need of replenishment. During the night of sleep, the bloodstream leaves much of the brain and goes to the digestive system. This explains the feeling of "fullness" upon awakening. Nature takes away much blood from the brain so it can rest and sleep. But this means that an undernourished brain has many millions of depleted, broken cells and tissues upon awakening. This *Morning Brain Booster* sends protein and Vitamin E in a unique *combination* to the billions of brain cells. Here, they speedily correct membrane stability, boost collagen (connective tissue) formation, send more nutrients to your tissues and send a stream of cellular rebuilding oxygen. Elizabeth M. says it's like "magic" for giving her youthful thinking in the morning. It's long-lasting, too!

Elizabeth M. is in her middle 60's but her mind is in its middle 20's, at the prime of life. She thinks clearly throughout the day and enjoys good health of her body and her mind, thanks to the cellular-cleansing-replication action of protein with Vitamin E in a Nature-created package, as found in whole grain foods and seeds, nuts and plant oils.

Having a young body is like having a good picture frame, but without a picture. Having a young brain in a young body is like having a complete picture of youth in one single package.

Let protein help you keep young in body and brain so you will enjoy the best that life has to offer. Protein, with its feeding and scrubbing power, will help give you the vim, the vigor and the will to live youthfully for the rest of your very long life!

IN REVIEW

1. Metabolized protein acts as food for your 16 billion brain cells.

2. Jeff H. doubled his thinking power by using protein for breakfast. Amino acids nourished the 3 "thinking" centers of his brain.

3. Lillian I. has "day long" mental vigor with protein snacks. She uses the Super-Protein Food for "dynamic" brain power.

4. Robert K. uses a simple Protein Punch for long-lasting brain vigor.

5. Whole-grain foods contain a unique combination of protein and Vitamin E that promote an anti-aging process of "brain cell washing." It is Nature's secret for a "forever young" brain.

6. Elizabeth M. follows a delicious 5-step "Brain Cleansing Program" with a *Morning Brain Booster* for amazing powers of "young thinking."

5

YOUR LIVER: SECRET OF CELLULAR REJUVENATION

Your liver holds the secret of cellular rejuvenation. A *healthy* liver is able to spark those processes that help produce a *healthy* body through constant tissue renewal. Your liver is unique as an organ of digestion and of excretion. But its prime value is that it must metabolize proteins and other substances needed for tissue and cell respiration. Your liver must be healthy because it sends amino acids throughout your body to all organs for the repair and regeneration of the billions of cells and tissues. *The healthier your liver, the healthier your body!*

Largest Body Gland. As the largest gland in your body, your liver weighs from 3 to 4 pounds. It is located in the upper right part of your stomach, just below your diaphragm. You can feel its outer border just beneath your ribs. Liver ducts connect with the gall bladder duct to form a common bile duct that empties into the duodenum (portion of your digestive system). Your liver acts as a filter as its own liver cells take up materials digested and absorbed from food. It holds an important "key" to cellular rejuvenation, and it is needed for basic health.

Millions of your liver cells, themselves, need protein. The liver uses protein to help build and rebuild your billions of body cells. It is important to see that you feed protein daily to your liver so that its columns of cells lining the numerous blood channels are healthy and strong. Because your liver synthesizes many amino acids for tissue building, uses amino acids in a form that gives you energy, produces substances that keep your bloodstream clear, and uses protein to detoxify drugs, poisons, chemicals and toxins from bacterial infections, you can appreciate the importance of having a healthy, protein-ized liver—for the sake of your liver and your body cells from top to toe.

SIX PROTEIN WAYS TO NOURISH YOUR LIVER
FOR CELLULAR REJUVENATION

Everyday healthful foods that are prime protein sources can serve a dual purpose. They can keep your liver in healthful regeneration and, secondly, they give your liver the working materials with which to rejuvenate the billions of body cells throughout your body. Here are six such all natural secrets of liver-and-body rejuvenation:

1. *Rebuilds Digestive System.* To help improve his digestive system, Earl O'H. takes a tasty and natural *Liver Elixir.* In a glass of tomato juice, he stirs one tablespoon of desiccated liver powder. Sold at most health stores, this is a dehydrated form of whole liver, but with no connective fat or gristle. Earl O'H. drinks *one* such *Liver Elixir* a day and feels that his entire digestive system has become rejuvenated.

Benefits: The vitamins in the tomato juice are taken by the proteins in the desiccated liver and help the body produce bile which is sent into the intestines to aid in the digestion of fats and other nutrients. The liver uses amino acids to act as transporters in receiving digested foods from the intestines. The liver also destroys a certain amount of proteins which cannot be fully utilized by the body economy. The liver needs the amino acids and vitamins-enzymes from the *Liver Elixir* to change substances into wastes for elimination through the kidneys. Earl O'H.'s simple *Liver Elixir* works to replenish the millions of cells and tissues that line the liver, where they act as reservoirs for the amino acids. He has a well-nourished liver and feels youthful all over because of this easy but effective program.

2. *Improves Storage Power of Liver.* Every day, Ruth N. plans on having one or more of these *complete* protein foods: lean meat, fish, dairy, eggs, seeds, nuts—in *combination* with grain foods such as whole grain breads or natural brown rice. This gives her liver a *balanced* protein of animal-plant foods and Ruth N. says that it boosts her overall body health. Her cheeks glow with health. Her skin is firm and smooth. Ruth N. is in her early 70's, yet she is amazingly alert, has clear eyes, and looks and feels the image of youth. She is basking in the rewards of a well-nourished liver with this protein combination plan.

Benefits: Taken in combination, a protein food from an animal and a plant source means that the amino acids are nourished together. The liver takes these *combined* amino acids to transform sugar and starch into glycogen to be stored within its cells. The liver uses these *combined* amino acids to replenish its own cells for better storage power. Stronger and repaired liver cells are able to convert glycogen into blood sugar to be sent to the muscles and other tissues, which creates a feeling of overall vitality and energy. *The healthier the liver cells, the better they can store and metabolize energy-producing glycogen.* Adequate protein-ization means well-nourished liver cells, the key to better storage of energy substances for better body-mind vitality.

3. *Boosts Iron Production.* Brenda O. worked in a warm dress shop, but she always had cold hands, cold feet, and a chilly feeling. Even when the heat was turned on "full blast," Brenda O. looked embarrassingly pitiful in a thick sweater, with a wan face and pinched cheeks. Iron foods offered little help. She soon discovered that for iron to be sent to her body organs to create warmth, she needed to regenerate her liver. Two simple foods in a simple combination gave her the youthful warmth she needed. She dubbed it a *Warm Up Potion.* Brenda O. would mix one tablespoon of Brewer's yeast in a glass of vegetable juice. A sprinkle of kelp powder (sea salt, sold at most health stores and food markets) was added for a piquant taste. She drank two of these *Warm Up Potions* daily. Within five days, she discarded her sweater and warm clothes. Her hands and feet were youthfully warm. She looked alert, healthy and happy. She radiated youthful warmth at the age of 64. All this, she feels, was a result of her *Warm Up Potion.*

Benefits: The Brewer's yeast is a Nature-created combination of all known B-complex vitamins and protein, with other nutrients, that join with the enzymes in the vegetable juice to alert-activate the liver cells. The liver uses these nutrients to take iron out of the system and store it within its own cells. Healthy liver cells can absorb many more nutrients for iron metabolism. The liver cells need protein as a spark plug to metabolize and release iron and fibrogen into the bloodstream to produce a feeling of warmth and good health. A protein-ized liver can help boost iron production throughout the body, and rejuvenate the bloodstream while it helps give a feeling of youthful warmth.

4. *Protein Protects Against Liver Cell Destruction.* Scott P. had a haggard complexion. He walked with a stooped gait. He had serious problems of acidosis that brought a feeling of "heartburn" right up in his throat. He would frequently regurgitate foods. Swallowing was difficult. A concerned co-worker told Scott P. that he was probably

protein-starved insofar as his liver was malfunctioning. He shared a secret with Scott P. Namely, eat a *Protein Sandwich* twice daily. Scott P. agreed, and while he had difficulty in eating at first, he soon responded and could eat with gusto. The amazingly simple secret was that the *Protein Sandwich* consisted of pure peanut butter (non-hydrogenated, as sold in health stores) with slices of cheddar cheese on whole grain bread. After *five* days, Scott P. perked up. He walked youthfully tall. His acidosis had subsided. His digestion boosted. He felt young again, thanks to this simple but effective protein plan. It was the delicious way to regenerate his liver, he found out.

Benefits: This *Protein Sandwich* contains as much as 60 grams of this needed liver food, almost the total minimum daily amount. The liver needs this nut-dairy-grain protein in a triple-barrel combination for nourishing its cells. This type of three-way protein rebuilds liver cells and enables it to screen out and protect against metallic poisons such as lead and mercury that enter the body through polluted air fallout. It is often unavoidable. Such internal pollution can cause an act of necrosis (death of liver cells) and reduce the effectiveness of this organ to keep the body healthy. To protect against necrosis, the liver cells need a simple combination of protein from three sources (for a balance). The *Protein Sandwich* offers this three-way source of protein: nut-dairy-grain for a completeness that is a secret of liver cellular restoration. It's the tasty way to protect your liver against pollution. It protects your body and your life, too!

5. *Protein Helps Improve Liver Filter Power.* Pauline C. said that she was a victim of lifelong "stomach gas." She felt bloated up, no matter how little she ate. She soon became malnourished because everything she ate reacted so adversely that she had to subsist on tea and toast and even this caused a back up sensation. She scoffed at the suggestion it could be her liver. But she listened to advice to make this easy dietary change. She discarded all of her bleached flour. Instead, she used *unbleached soy flour* for baking. She used it for muffins, for homemade bread, in meat loaves, and in just about any recipe calling for flour. This easy change produced a miracle of healing. Soon, she was able to eat and enjoy most foods with gusto. Gone was the "stomach gas" and gone was the "bloated" feeling. She uses soy flour daily. She calls it her *Soy Pro Plan.*

Benefits: Just *one cup* of soy flour can give you close to over 80 grams of protein—*more* than the basic minimum required for health. The protein in soy flour is unique in that it is used by the liver cells (which act as filters) for self-renewal. These liver cells use soy flour protein to neutralize certain poisons which are manufactured by the body or otherwise received into it. For example, your liver cells use protein to neutralize the nicotine of tobacco smoke and pollution and prepare it for elimination through the kidneys. Your liver cells need protein to help neutralize some of the toxic wastes of refined foods. Your liver cells turn them into neutral or harmless compounds for release. Certain intestinal wastes that are not eliminated by the colon are transported into your liver. Here, your cells need to detoxify these wastes for eventual release through the kidneys. A simple *Soy Pro Plan* gives your liver cells a treasure of amino acids needed for its own self-regeneration. Then the liver cells can effectively perform this "liver filter power" that will help keep your body sparkling clean. Soy flour is about 90% pure protein and needed as protein food by your liver cells. It's the tasty way to take your protein!

6. *Maintains Youthful Cellular Balance.* Eleanor F. ate very little because she had to watch her figure. She gained all too easily. But this reduced food intake made her undernourished. In particular, her liver cells were so starved for protein, she reacted with feelings of biliousness, headache, fatigue, a general feeling of malaise. She looked for a protein food that would be nourishing but low fattening. She found it on her grocer's shelf in the form of *low-fat dry skim milk powder.* But she disliked milk, even if it was low fat. So she made herself an easy protein meal: *Protein Soup.* To any homemade soup, whether vegetable, meat, fish, bean, pea or broth, all she did was add four heaping tablespoons of low-fat dry skim milk, stir vigourously, and then eat with a slice or two of whole grain bread. This easy *Protein Soup* program, followed for just nine days, worked wonders with Eleanor F. She remained slim but not undernourished. More importantly, her liver was now nourished so her biliousness subsided, her headache spells ended, her fatigue turned into youthful energy and she radiated the joy of living—thanks to a nourished liver!

Benefits: Dry skim milk powder, whether low fat or not, is an excellent source of dairy protein. This type of protein is easy for liver absorption. The liver cells need dairy protein because it is mild and can correct problems of inflamed bile production. Liver cells use protein to create an alkaline type of bile which acts as an antiseptic or bactericide, and destroys germs and wastes. The liver cells transform protein into a form of amino acid used by the bile to neutralize excess acidity in the digestive system. (Many folks troubled with ulcers or bile distress are advised to eat and drink dairy food because its protein is especially beneficial to the liver cells.) To get the most benefits from the protein in dry skim milk powder, use it for making a beverage, in soups, or anywhere else in cooking. It offers your liver cells a unique and beneficial form of protein for self-regeneration and protection against cellular destruction.

15 PROTEIN FOODS TO HELP REJUVENATE YOUR LIVER CELLS AND TISSUES

Here is a listing of 15 good protein foods that can give your liver the needed ingredients for cellular-tissue rejuvenation. Plan to use as many of them as possible in *combination* for better body health.

1. Eggs

Plan on 2 or 3 eggs weekly. An egg is *complete* protein. If you are cautious about cholesterol in the egg yolk, then eat egg white daily (it's pure protein and has *no* cholesterol) in the form of scrambled eggs with whole grain bread. Try eggs in custards, omelets and French toast. This gives you good protein nourishment.

2. Milk

Plan on 2 to 4 glasses daily. If you are on a low-fat, low-cholesterol program, just switch to skim milk. Enjoy yogurt or buttermilk. Drink milk plain or make custards and soups with milk. Use dry skim-milk powder in soups, salads, baked goods and loaves, as well as casseroles. You will feed your liver and your body cells a tremendous treasure of amino acids with balanced vitamins and minerals.

3. Liver

Yes, the best way to regenerate your liver is by eating liver. Try lamb liver, beef liver, chicken liver and calves' liver. You feed yourself a superior protein with vitamins, minerals and other nutrients. Try broiled or baked liver as a luncheon dish with whole grain bread. Twice a week will be feeding your body a steady supply of cellular rebuilding protein.

4. Brewer's Yeast

One of the best sources of plant proteins available, in a Nature-created balance of vitamins and minerals. Either Brewer's yeast or torula yeasts are good. A concentrated form of protein. Use it in beverages, mix in baked dishes, casseroles, in soups and just about anywhere. Try a half teaspoon in a bowl of whole grain cereal. The feeling is one of invigoration and nourishment as the protein goes to work almost immediately in liver cellular regeneration.

5. Soy Flour

Sold at most health stores and many supermarkets, it should replace all other flours. Use it for breakfast pancakes, waffles, muffins, rolls and bread. You may even use it in puddings, custards and many desserts. It is almost total protein and a powerhouse of amino acids for liver and body renewal.

6. Cheeses

Farmer cheese, cottage cheese and cheddar cheese are just a few of the available cheeses that offer soothing protein for liver regeneration. Cheese can be enjoyed daily. Use fresh fruit slices for a double-action benefit because the Vitamin C in citrus fruits can speedily metabolize the protein in the cottage or other cheeses. A luncheon platter of assorted cheeses and fruits with melba toast offers a rich assortment of different proteins for body health.

7. Meat, Fish, Poultry

Excellent sources of protein, iron, minerals and vitamins that complement one another in their effort to renew the billions of body cells and the liver tissues. Plan on eating two or three of these foods during the week. Trim off visible fat. Eat with a large raw fruit or vegetable salad for better vitamin-enzyme metabolism of the protein.

8. Soups

Any simple soup can be transformed into "protein powerhouses" by adding Brewer's yeast, skim milk powder, diced or shredded peanuts, any types of seeds and nuts, or even some soy flour. Make a rich protein broth out of cooked vegetables, adding some powdered milk and a half teaspoon of Brewer's yeast. Flavor with kelp. You'll have an amazingly effective supply of cellular nutrients.

9. Vegetables

While comparatively lower in protein, vegetables offer a prime source of enzymes, vitamins, minerals and roughage. These are needed to work with protein for liver regeneration. Without these nutrients, protein could not be metabolized. Select richly colored vegetables. Eat them raw, as much and as often as possible. Cook only if they must be cooked. Vegetables offer potassium which is used by the liver for better amino acid metabolism.

10. Fruits

A major source of enzymes and Vitamin C needed by the liver to regenerate its cells and then to transform protein into amino acids for tissue respiration. A *combination* of any protein food with fruit gives it a super-action power and makes its amino acids more readily available for utilization by the billions of body and liver cells. Use fresh fruits in season. Otherwise, try frozen fruits. Canned fruits packed in natural waters and *not* in sugary syrup (destructive to liver cells) are also satisfactory. But the emphasis is on fresh fruits for optimum vitamin-enzyme power.

11. Whole Grains

Breads and cereals made from fresh, stone-ground grains and flours are prime sources of protein and B-complex vitamins, with needed Vitamin E. Remember to use wheat germ, too. Use whole grain breads, cereals, hotcakes, muffins, waffles and use them as "fillers" in casseroles, soups, meat loaves. Grain protein is easily used by the liver for its self-renewal processes.

12. Vegetable Oils

Use any cold-pressed vegetable oils made from seeds, nuts, or plants of any sort. They offer good amounts of Vitamin E with protein in a combination that helps "wash" out cellular litter from your liver. You'll get this same benefit from non-hydrogenated nut butters. Use oils as a flavoring. Use nut butters as sandwiches. A top-notch way to cleanse your liver through the action of Vitamin E and protein, and then to have the liver cells regenerated through the metabolized amino acids. Use daily.

13. Lecithin

A soybean derivative that is said to help prevent fat accumulation in the liver. Lecithin protein acts as a buffer zone to see that liver cells do not become too "fatty", and keeps them scrubbed clean. Lecithin, sold at most health stores and many special food shops, is also available in liquid form, capsules, powder, granules and tablets. Use lecithin sprinkled over soups, salads, in meats and even on dairy products. It helps cleanse your liver while its protein helps rebuild cells and tissues.

14. Kelp or Sea Salt

An edible seaweed type of salt substitute that is a prime source of ocean minerals such as iodine, calcium and potassium, as well as many vitamins and protein. This *combination* offers an "all purpose" nourishment for your liver cells and "feeds" them so they can self-reproduce and improve the health of your liver—and body, too! Use kelp as a salt substitute!

15. Raw Salads

Whether fruit or vegetable, raw salads should be eaten daily. With some cheese slices and fresh fruits, a sprinkle of Brewer's yeast, almonds, nuts and seeds, you have a tremendous liver-feeding source of protein, vitamins, enzymes and minerals. A raw salad has a special benefit: *it contains those nutrients required to transform protein into cell-rebuilding amino acids.* Without the substances in raw salads, protein would lie undigested in the system! A raw fruit or vegetable salad is a delicious "must" in your liver cell-tissue rejuvenation program.

Your liver (and body) cells and tissues can rebuild at top notch speed only when protein and other nutrients are available in plentiful supplies.

A healthy liver can be your ticket to the road to perpetual youth! Protein acts as your ticket taker as you journey forth on the way to renewed youth through a cellularly-renewed liver!

IMPORTANT POINTS

1. A healthy liver offers hope for body youth and health. Note the 6 ways to use protein for cellular rejuvenation of the liver.

2. Earl O'H. takes a natural *Liver Elixir* for digestive rejuvenation.

3. Ruth N. uses an easy *combination* of protein foods to improve the storage power of her liver.

4. Brenda O. uses a *Warm Up Potion* that so regenerates her liver, that the iron in her body is better metabolized and she enjoys warm hands and feet at a young-looking 64.

5. Scott P. eats a *Protein Sandwich* containing three protein foods that work miracles in boosting his digestion, soothing his acidosis through liver regeneration.

6. Pauline C. improves her liver filter power and corrects "stomach gas" distress on an easy Soy Pro Plan.

7. Eleanor F. uses a *Protein Soup* to nourish her liver cells; she has overcome problems of biliousness, headache, fatigue and general malaise, thanks to this easy and tasty cellular balance program.

8. Feed your billions of body and liver cells a daily supply of protein, along with other nutrients, by following the easy 15-step program outlined. All foods are available anywhere at modest cost. Many are in your pantry now. It is not what food you have but *how you eat it* that helps regenerate your liver and give you decades of extra health-packed years of youthful living.

6

HOW PROTEIN HELPS YOUR INTERNAL ORGANS FEEL "FOREVER YOUNG"

Nature wants to help you feel "forever young" in body and in mind. Gerontologists (specialists in the study of aging cells) have discovered that the secret "cure" for aging lies in the proper intake of the right kind of protein. This miracle cell rebuilder is able to protect against aging by performing some of these basic functions:

1. *Tissue Rejuvenation.* Protein works with Vitamin C to manufacture collagen, the substance that is the major ingredient in skin and other connective tissues. Protein helps the body make an adequate daily supply of this needed cell-tissue builder. Protein uses collagen for rebuilding and rejuvenating billions of body cells and tissues.

2. *Creates Immunity.* Protein is used to nourish the body so its immune network will respond to protect you against infections and other invasions. Protein helps the body produce immune cells and antibodies to act as a shield against harmful viruses and corrosive antagonists.

3. *Controls Formation of Tarry Substance.* Protein is needed to control the formation of *lipofuscin,* a brown, tarry substance that accumulates within the cells, crowding out the more useful cellular rebuilding substances. Protein appears to inhibit the formation of this *lipofuscin,* a direct effect of internal cellular aging.

4. *Maintains First Class Cellular Health.* Adequate protein enables body cells to produce more vigorous substances that bring about self-renewal. A protein inadequacy leads to *weakness,* whereby the cells produce defective substances and tissue breakdown is the inevitable risk.

5. *Creates Internal Harmony.* Sufficient protein helps "control" cells, gives them discipline to protect against becoming wild-growing or self-destructive. Protein is required to maintain such an internal discipline.

NATURE'S PLAN FOR PERPETUAL YOUTH

With adequate nutritional nourishment, protein can help give you a look and feel of perpetual youth. In general, any given function in any given person declines by 1 percent a year after the age of 30. Draw a sloping line and it will reach the bottom at age 120. Nature has a plan to help you reach 120, but with your internal organs feeling healthy so that you can enjoy the best that life has to offer. With the right kind of protein, this can become a "possible dream."

THE MEAT PROTEIN THAT HELPS TAKE YEARS OFF YOUR FACE AND BODY

Vera Q. is "over 50" and has three grandchildren. Yet she displays the energy and youthful appearance of a woman half her age. She is active at sports. She plays tennis and frequently beats her younger opponents. She has corrected her problems of indigestion, she has solved her problems of "stomach burning" and distressful colitis. She has given her internal organs the "safe meat protein" that is considered a miracle food for the organs of the digestive system, the key to self-perpetuating youth.

Eats Turkey Meat Regularly. Vera Q. is concerned about cholesterol clogging of her cells and tissues. Since nearly all animal foods contain cholesterol and protein, she is selective in her meat foods. Vera Q. has discovered that turkey meat offers two unique cell-tissue rebuilding secrets:

1. Turkey is highest of all cooked meats in protein.
2. Turkey is lowest of all animal foods in cholesterol.

Vera Q. satisfies her meat quota by eating turkey throughout the week. She eats other meats on occasion, but turkey is the "star" in her cell-rebuilding program and she radiates youthful health. Her internal organs are benefiting by turkey protein which forms nitrogen-containing substances essential to body rejuvenation.

Rebuilds Internal Organs. Vera Q. looks and feels youthful because metabolized turkey meat protein is constantly being broken down and is resynthesized to rebuilding the billions of cells and tissues in the liver, gall bladder, colon, stomach, pancreas, intestine and connected organs. Vera Q. benefits because turkey meat has *complete* protein containing *all* essential amino acids needed for cellular rejuvenation—but with *much, much less cholesterol.* This is the lipid (fatty substance) that may collect on the inner walls of the digestive system's blood vessels just as rust collects on the inside of an iron pipe.

[Excessive cholesterol buildup may partially cut off the flow of blood. This is the forerunner of hardening of the arteries (arteriosclerosis) and dreaded coronary heart disease.]

Vera Q. eats a wholesome diet consisting of large amounts of fresh fruits and vegetables, whole grain breads and cereals, two eggs per week, skim milk dairy foods, occasional beef and liver, and regular amounts of turkey meat. She radiates youthful health as a reward!

Extra Bonus: Turkey is lowest in cholesterol of all popular meats. Its low fat content is classed among the soft-type fats (unsaturated), the kind that does not increase the blood cholesterol level.

Easy to Digest. Your billions of cells and tissues in your digestive organ welcome turkey protein since it is considered very easy to digest and offers high-quality protein to build the needed substances to promote internal rejuvenation.

Top Notch Protein. The breast meat of cooked turkey ranks higher in protein than any of the other prepared meats. Turkey legs and thighs rank only slightly under breast meat in protein content.

Eat Turkey Regularly. Rebuilding your internal organs calls for "easy" metabolism to spare extra effort. Turkey protein offers this gentle but effective way of soothing while healing, and then promoting a form of natural rebuilding of your tissues and organs.

A SIMPLE PROTEIN PLAN FOR REBUILDING YOUR INTERNAL ORGANS

Gerontologists note that there is a unique cell rebuilding process occurring on the principle of "like heals like." Namely, to rejuvenate your liver, eat liver. To rejuvenate your heart, eat heart. To rejuvenate your skin, eat whole meat products. This is a simple protein plan that should help to rebuild your internal organs. Here are the benefits of such a simple protein plan:

1. "Like heals like" protein foods can travel as "whole cells" throughout your body to the internal organ needing repair and cellular regeneration. A digested liver protein offers amino acids to the body's own liver to help repair any damage and restore cell reserves. It is a form of cellular healing that may offer a clue to rebuilding specific parts of the internal system.

**Nutrient composition
of cooked turkey meat
in relation to other cooked,
boneless cuts of meat
and other protein foods**

Food	Amt.	Weight mg	Cholesterol* mg	Protein mg	Calories (food energy)	Riboflavin mg	Niacin mg
Turkey							
white	3½ oz.	100	8 to 15	33	176	0.14	11.1
dark	3½ oz.	100	16 to 26	30	203	0.23	4.2
Chicken (white)	3½ oz.	100	60 to 90	32	182	.1	11.8
Chicken (dark)	3½ oz.	100	60 to 90	29	184	0.2	5.3
Veal (chuck)	3½ oz.	100	90	28	235	0.3	6.4
Beef (round)	3½ oz.	100	125	29	261	0.2	5.6
Pork (loin)	3½ oz.	100	70 to 105	25	391	0.3	5.8
Lamb (rib chops)	3½ oz.	100	70	20	407	0.2	4.6
Salmon, canned Chinook	3½ oz.	100		20	210	0.14	7.3
Eggs	2 med.	100	550	13	163	0.3	0.1
Shrimp, canned	1⅔ oz.	50	63	12	58	tr.	.9
Cheese							
cottage, creamed	⅜ cup	100	15	14	106	0.3	0.1
cheddar	1 oz.	28	28	7	112	0.1	tr.
Butter	1 pat.(¼oz.)	7	18	tr.	50		
Milk							
whole	1 cup	244	28	8.5	160	0.4	0.2
skim	1 cup	244		8.8	88		0.4

Agric. Handbook No. 8: Composition of Foods. 1963

*Cholesterol values from Mitchell
Meat and Poultry statitics for uncooked portions

Chart 2

Chart reprinted through courtesy of *Natural Health Bulletin,* Parker Publishing Company, Inc., West Nyack, N.Y. 10994. Vol. 3, No. 22, October 29, 1973.

2. Organ meat protein becomes metabolized by the body and then offers the *mitochondria* (a compound found in the cell and a prime source of enzymes and protein to be used as the basis for cellular replication) to the internal organ for cellular rebuilding. Organ meat protein, in the form of *mitochondria,* involves energy conversion within the cells to promote regeneration.

A simple protein plan would call for eating liver for helping the liver regenerate, heart for helping the heart cells replicate, kidney for helping the kidneys undergo rejuvenation, pancreas for helping this insulin-producing gland become stronger and more effective through cell-tissue respiration. The "like heals like" discovery can be the secret for self-perpetuating internal organs through organ meat proteins.

HOW BUTTERMILK PROTEIN IS NOURISHING FOR THE KIDNEYS

Benjamin B. has long been troubled by less-than-adequate kidney health. He has "sour taste" and an "acid stomach" as well. He complains of a feeling of heaviness even after a little bit of eating.

Simple Food Nourishes Kidneys. Benjamin B. has discovered that buttermilk protein can become nourishing for the kidneys by a unique action of cell-rebuilding. Buttermilk contains lactic acid, a protein-like substance formed by the fermentation of lactose, a native substance in milk. This buttermilk protein is in the form of a fine curd, which permits buttermilk to be more speedily digested than ordinary milk. To help keep his kidneys healthy, Benjamin B. drinks two or three glasses of buttermilk daily. As a result, his kidneys are nourished by this type of protein. His sour taste is gone. His acid stomach is soothed.

Benjamin B. drinks one glass of this delicious *Alka-Protein Nog* daily to help content his digestive system, and to help nourish his millions of kidney cells. How to make it:

Alka-Protein Nog

Combine 1½ cups chilled buttermilk, 1 banana, sliced, 1 tablespoon honey, a dash of cinnamon. Beat well with a rotary or electric beater, or use an electric blender.

Secret of Kidney Rebuilding Powers of Alka-Protein Nog: This is a high alkaline food which is soothing for "acid stomach" often traced to kidney disintegration. The carbohydrates of the banana have a protein-sparing action on the protein of the buttermilk. This soothing cell-rebuilding action is maximally effective with buttermilk protein, and fruit carbohydrates are introduced to the digestive system simultaneously. The banana nutrients help the buttermilk protein become better metabolized for kidney cell rejuvenation. This very low sodium content makes the *Alka-Protein Nog* even more soothing to the kidney cells. A simple program, yet the kidney and body cells are rewarded with a treasure of protein replenishment.

THE FERMENTED MILK FOOD THAT PUTS YOUTH INTO YOUR DIGESTIVE SYSTEM

Donald O'B. is always "on the go" and this means his body is subjected to a great deal of wear and tear. His constant state of tension has given him a feeling of being "bloated", and also problems of constipation and colitis.

Vicious Cycle Destroyed Millions of Cells. Troubled by constipation, Donald O'B. took harsh chemical laxatives which destroyed millions of his digestive-intestinal canal cells. But then he had problems of colitis as well as diarrhea, which further depleted his body of needed nutrients and cells. Donald O'B. was pale, wan and shaky. His digestive system was tortured by the vicious cycle that led to under-nutrition and premature aging.

Old World Food Produces a New Life for a New Digestive System. Donald O'B. disliked dairy foods but he was introduced to yogurt which had a delectable taste. More important, the old world food (a staple in middle Europe and Central Asia where yogurt eaters reportedly live well beyond the 100-year span, and have little or no digestive troubles) soothed, healed and regenerated his digestive-intestinal system. Eating yogurt daily, especially *before* a meal, gave him such a youthful feeling, his digestion improved and he could actually throw away his laxatives. Donald O'B. experienced complete recovery from his constipation-colitis syndrome and he no longer felt bloated. His skin firmed up. His energy increased. His body responded with more youthful alertness. He gives credit to his eating of yogurt at least thrice daily.

How Yogurt Replenishes Cells and Tissues of Digestive System. The food you eat contributes to your youthful cellular respiration.

But the nutrients you *assimilate* offer the secret key to overall body rejuvenation. Regular milk is only 32% digested after being an hour in your digestive tract. But 91% of yogurt is digested within the same time. Its protein is superior because it has been increased by the act of *fermentation.* This process creates the formation of *lactobacillus Bulgaricus,* which is then used by *casein,* the unique milk protein, to destroy putrescent bacteria in your digestive-intestinal system. In effect, yogurt protein is used to cleanse the intestinal flora pattern of your digestive system, and protect against undesirable organisms. Yogurt protein, being of a *fermented* kind, has super-acting power upon the digestive-intestinal canal, and is able to rebuild and regenerate the millions of cells and tissues needed for youthful health.

Yogurt protein is metabolized almost three times as quickly as ordinary milk protein. Yogurt protein offers you "lightning swift" assimilation so that your digestive-intestinal system can be repaired very quickly. This makes it an ideal food to be enjoyed several times daily for day-long cell-tissue respiration.

Yogurt Protein + Fruit Vitamin = Super-Cell Repair. You can double or triple the cell replication power of yogurt protein by adding several slices of fresh seasonal fruit to your portion. *Benefits:* Yogurt protein joins with the natural ascorbic acid (Vitamin C) of fruit to promote rebuilding of your billions of digestive cells by creating collagen to create new cellular walls that have been broken off. Without Vitamin C, protein would remain ineffective. But when your body receives this *combination* of fermented milk protein in yogurt with Vitamin C in citrus or other raw fruits, your billions of cells and tissues are super-nourished with collagen, the key to cellular rejuvenation. *It is the natural way to extend the prime of your digestive life.*

HOW TO DRINK YOUR PROTEIN FOR INTERNAL CELLULAR REBIRTH

Stella A. was troubled with irregularity. Her discomforts were manifested by symptomatic distress of related hepatitis and cirrhosis of the liver.

(These liver ailments are traced to defective cell rebirth, during the natural process of metabolism. There is a hardening of the liver, often accompanied by destruction of its cells, for which a "leathery" or defective type of tissue is substituted. It is often traced to a protein deficiency that renders the body susceptible to such a health decline.)

Looks Wan, Pale; Feels Cold, Aged. Stella's poor digestive powers brought about protein deficiency which denied her liver and con-necting organs their needed nourishment. Cellular necrosis made her look worn and pale; improper circulation and a poor filtering service made her feel very cold and aged. Stella A. had such a poor appetite, she could not bring herself to eat the nourishing foods required for a natural liver regeneration. So she had to *drink* her protein. She did this by a simple and amazingly powerful beverage. Stella A. responded so that her liver regenerated itself and put color into her cheeks, smoothed out her aging lines, boosted a healthful appetite and put warmth into her fingers and hands. Stella A. was also "regular" because of this simple but powerful protein-regenerating drink. It was a miracle of protein power. Here is how to make it by using three everyday items found in almost any food market:

Cell Youth Elixir. In a glass of fruit or vegetable juice, dissolve two heaping tablespoons of *unflavored* gelatin. Add two tablespoons of Brewer's yeast. Stir vigorously. When thoroughly blended, drink slowly.

Secret Benefits: The secret cellular rejuvenating power in the *Cell Youth Elixir* lies in its *unflavored* gelatin, one of the most potent sources of "instant" protein. Gelatin, itself, is made by boiling the high-protein bones and skin of animals, and then reducing the result to a fine powder. Your body contains *collagen,* a protein component of connective tissue that is almost *identical* in content to the gelatin powder in your *Cell Youth Elixir.*

The protein of this drink goes into your digestive system where it is metabolized and then sent to your skin, blood vessels, bones, tendons, cartilage and connective tissues.

The vitamins from the juice work upon the protein of gelatin to create this building and rebuilding amino acid feeding of your internal organs. The Brewer's yeast contains several amino acids missing from gelatin. So the *Cell Youth Elixir* gives you *complete* amino acid supply in a form that is just about identical to that of your body collagen. When you drink your natural *Cell Youth Elixir,* you are giving your organs *collagen,* to replace that which is lost during day-to-day aging. It is the fulfillment of the "like heals like" secret of self-rejuvenation. Feed your organs that which they lose and this *replacement* offers a protective bulwark against unnecessary aging.

HOW PROTEIN HELPS GIVE YOU A YOUTHFUL "BREATH OF LIFE"

Stephen T. was always wheezing, gasping and choking for air. The slightest bit of dust would send him into a coughing and sneezing fit. Often, he would run to the window and gulp in air. Since he lived in a highly polluted region, this worsened his problem. Stephen T. was aging rapidly because he was being deprived of the "breath of life."

How Protein Snacks and Nibbles Gave His Lungs a New Lease on Life. Because he was careful about his weight, Stephen T. ate sparsely and this sometimes creates a protein deficiency. But he was told that he could protein-ize his respiratory system by ordinary "snacks and nibbles." Here is his simple program:

Throughout the day, Stephen T. nibbles on cheese squares and whole grain bread. He munches raw seeds and nuts, chewing them *thoroughly* so his digestive enzymes can use the metabolized amino acids for respiratory and internal organ regeneration.

For extra-powerful protein metabolism, he often soaks nuts in milk or pineapple juice for twenty-four hours in his refrigerator. The enzymes and vitamins in the liquid pre-digest the nuts so that when later eaten (the liquid is sipped, too), their amino acid content is released and made speedily available for the billions of waiting cells and tissues. It is the "instant" way to nourish the cells.

Stephen T. adds skim milk powder to soups, casseroles, puddings, gelatin dishes, meat or fish loaves, and to gravies and broths.

Breakfast Protein Booster: Stephen T. would make a simple breakfast omelet to which he would add skim milk powder and some wheat germ. This omelet proved to be a protein booster of enormous benefit. Such a high-protein breakfast immediately sent amino acids to the cell-starved respiratory system. Collagen soon rebuilt his lungs so that his coughing spells eased. His nighttime spasms also subsided. He could breathe easier. He felt as if he had been a drowning man, just coming out of water for beautiful, delicious fresh air. Protein, whether snacked or nibbled, or eaten, gave his respiratory system the needed nutrients for cellular rebuilding.

How Protein Boosts Breathing Powers. A microscopic cross-section view of the organs of respiration would show millions upon millions of tiny air cells and sacs. These include the bronchial tubes, the nasal passages, the pharynx, trachea, the lungs themselves, and the blood vessels that pick up oxygen. The tiny bronchioles, air sacs (alveoli) of the lungs, and the muscles of respiration all need a never-ending supply of collagen for the regeneration of their billions of tiny cells and tissues. They use protein to introduce oxygen from the air, propel it into the bloodstream, nourish the blood cells, and maintain a healthful ventilation of the entire respiratory system. Protein acts as a natural propellant. Protein nourishes and builds these billions of cells. Since breathing is a never-ending process, many millions of cells are broken and destroyed during the effort. Protein must be made available at all times and in adequate amounts so that these "breathing cells" can be repaired and rejuvenated. A deficiency may interfere with, reduce and choke off the very breath of life!

10 PROTEIN FEEDING PROGRAMS TO NOURISH YOUR BREATHING SYSTEM

Here are some very easy ways to protein-ize your respiratory system, and to nourish your lungs so that you can breathe easier and better.

1. Snack on nuts, seeds, whole grain bread squares with some fresh fruit slices.

2. Protein-fortify most meals with skim milk powder.

3. Dissolve unflavored gelatin in a glass of skim milk and a squeeze of lemon juice for a tangy and super-treasure of cell-building collagen.

4. For a sore throat or persistent coughing problem, give your cells and tissues a treasure of protein by drinking a cup of herb tea to which you have added one tablespoon of Brewer's yeast powder.

5. A cup of cottage cheese with wheat germ and fruit slices.

6. Sprinkle shredded or powdered nuts and seeds over any dessert.

7. A chopped turkey salad, sprinkled with unflavored gelatin powder.

8. A chopped chicken salad, sprinkled with any diced nuts or seeds.

9. A whole grain breakfast cereal with skim milk, sun-dried raisins, and a sprinkle of Brewer's yeast.

10. Dissolve one tablespoon of desiccated liver in fruit juice. Drink throughout the day.

Better Protein, Better Breathing, Better Respiration. Protein food sliding down your throat during the swallowing process should be strong enough so that the amino acids are used by the billions of respiratory cells and tissues lining your frontal chest cavity. Aim for a *complete* amino acid pattern with some of the 10 protein programs outlined above. This complete pattern gives your respiratory system a balance of needed amino acids for regeneration of these cells and tissues. Keep them in optimum regeneration and you will be able to enjoy better breathing and better respiration. It's your very "breath of life" and youth, too.

Keep your internal organs well-nourished and you can help rebuild your body from the inside to the outside. Protein is Nature's plan for a feeling of perpetual youth. If you're young *inside*, you'll be young *outside!*

IMPORTANT POINTS

1. Protein offers five internal rejuvenating benefits to help control the so-called aging process.

2. Vera Q. enjoys turkey, a tasty meat protein that helps take years off the face and body. Its amino acid pattern is complete and has solved her problems of colitis and digestive distress. It is the *lowest* known cholesterol meat with the *highest* amount of protein in one package! No other meat is like it!

3. The "like heals like" principle enables organ meat protein to act specifically upon certain internal organs. Liver helps liver cells, heart helps heart cells, kidney helps kidney cells, pancreas helps pancreas cells. It offers hope for specific organ regeneration through protein foods.

4. Benjamin B. mixed 4 everyday ingredients to form an delicious and all-natural *Alka-Protein Nog*. It sends a balanced amount of protein throughout his system to help nourish his kidneys, ease "acid stomach", and soothe his entire internal organism.

5. Troubled with irregularity, Donald O'B. uses yogurt as a healthful way to offer protein to his digestive system. Combine yogurt with fruit for a unique super-cell repair action.

6. Stella A. soothes and nourishes her liver, puts warmth and color to her cheeks, by mixing together 3 everyday ingredients in the form of a *Cell Youth Elixir*. She drinks her way to better internal youth and health.

7. Stephen T. nourishes his protein-deficient respiratory system by snacking and nibbling on easy, everyday foods. He uses skim milk

powder for a tasty way to send protein to the billions of his respiratory cells and tissues. It helps him overcome problems of choking, allergic sensitivity, breathing distress, frequent sneezing and coughing. Protein has given him a new "breath of life."

8. Supercharge your internal organism with protein in any (or all) of the tastefully healthy 10 protein feeding programs to nourish your breathing system.

7

HOW AMINO ACIDS HELP
LOOSEN UP ARTHRITIC STIFFNESS

Arthritis may be described as an error in metabolism whereby the components within the bone structure do not renew themselves properly. The symptoms of such a' malformation include inflammation of the joint as well as an occasional dull ache to more serious developments such as stiffness and gnarled limbs. It is considered Nature's warning that the millions of cells which make up the site of the malformation are in need of *biological treatment*. This is one of the most recent discoveries as to the causes and possible healings of the fastest-growing of all degenerative disorders.

CROSS-LINKAGE—METABOLISM ERROR THAT CAUSES ARTHRITIS

Misformed or missynthesized proteins can cause adverse cellular reactions that lead to an error in metabolism which creates *cross-linkage,* the clue to arthritis development. These misformed proteins are not easily or properly integrated into cells because of stereochemical misfit and abnormal conformation alterations. They erroneously form into "twisted" cells that are known as cross-linkages. The symptoms cause inflammation of the synovial membrane and there is a general decline of health. Cross-linkage is the metabolic error that is traced to an improper protein intake. More important, it is traced to an *imbalanced* amino acid intake wherein there is a breakdown and an erroneous healing of the cells located within and around the bone structure.

The Nutritional Error That Creates Cross-Linkage. Basically, proteins are long chains of amino acids that are linked together. Adequate nutrition helps these long chains mend and knit together so that the cell structure becomes a strong and cohesive foundation for the bone or the joint or the muscle. But an *error* caused by inadequate nutrition or *imbalanced* protein can cause a malformation of the cell structure and cross-linkage takes place. This is the nutritional error that is the predisposing factor in arthritis distress.

Cross-linked cells may be visualized as small rods with two hooks which connect together two huge molecules. A nutritional error incorrectly ties the cells together so as to hinder the free moving range of the organism. It is comparable to having chains put on the limbs. Efficiency is impeded. The entire organism is slowed down unless a method is found to remove the "chains" from the cells. One symptomatic reaction is that of arthritis.

HOW PROTEIN BALANCE HELPS REMOVE "CHAINS" FROM CELLS

The protein molecules are often parallel and close together. If there is an imbalance or a limitation on protein intake, then the long molecules struggle to rebuild and knit together. Without the needed amino acids, the deprived molecules are forced to form "covalent" or non-free bonds which are known as cross-links. These non-free chain links are very strong and difficult to remove. If they continue to form because of malnourishment, then the properties of the molecules are even more drastically changed and cause health decline. These cross-link "chains" tend to "strangle" many healthy cells, which further deteriorates the structure and worsens the condition. There are good indications that *balanced protein intake,* with *optimum nutritional care,* will help send amino acids to deactivate or slow down the process whereby cross-linkage occurs.

Amino Acids Promote DNA-RNA Rescue Help. Digested proteins become amino acids through the help of vitamins-minerals-enzymes in many foods. This combination transforms protein into *usable* amino acids, which then initiate the manufacture of the nucleic acids known as DNA-RNA. Amino acids promote the flow of such cell-feeding substances which travel directly to the site where protein is needed.

Amino acids help create formation of these nucleic acids, and then send them as "messengers" so that they can help build and rebuild the billions of cells and tissues in need of repair. A deficiency or imbalance of protein means that the DNA-RNA cell-feeding substances cannot be sent to the "damage" site in quick or sufficient enough amount. Therefore, the cells become "cannibals" and seize upon any available protein for self-repair. This may often create a missynthesized type of protein which is not body characteristic. It is a "mistake" that leads to cross-linkage or "strangulation," and such distress symptoms as arthritic unrest.

An *adequate* and *balanced* variety of all amino acids means a "supply pool" of the materials needed by the cells for corrective rebuilding.

Why Eggs Offer Hope for Cellular Rebuilding. One particular balanced amino acid food is that of the egg. Located within the egg is a *complete* amino acid pattern that offers the cells the needed working materials for replication. *Benefit:* The egg contains a high supply of sulfur amino acids *(cystein, cystine, methionine)* which are recognized as being extremely important for cells and tissues. It is the *one* food that can offer hope for proper cell rebuilding and protection against arthritis-reacting "cross linkage" in a balanced food program.

Cholesterol-Free Plan. An egg yolk is the single most concentrated source of cholesterol. (One yolk contains about 250 milligrams of cholesterol. The recommended daily intake is set at about 300 milligrams.) To protect against cholesterol, you may eat just two or three eggs *per week.* This includes the number you use in cooking and baking. But you can have much of the needed sulfur amino acids in eggs without the cholesterol. *Eat the egg whites!* This will help give you a *complete* amino acid pattern containing *the three basic amino acids* that are known for being specific foods for healthful cellular rebuilding. They are free of cholesterol, and act as a protective program against arthritis degeneration through malformation of tissues.

Simple Program: In a balanced food plan, use egg whites scrambled in vegetable oil and sprinkled with wheat germ for breakfast. Add several slices of fruit so the enzymes can better metabolize the needed amino acids into a form that will be used as nourishment by your cells and tissues. Use egg whites in batter for whole grain pancakes, waffles, breads, rolls and muffins. This gives your body a mixture of different proteins (wheat and whole grains have good plant proteins) that work with egg white protein to provide a *balance* of cellular rebuilding. For a main meal, add stiffly beaten egg white as a "binder" to meat loaves or casseroles. You may also blend stiffly beaten egg white with honey and use as a topping over a fruit dessert. Use the same topping over gelatin for a super-protein energizer for your billions of body and bone tissue cells.

Rewards: The use of sulfur amino acids with other amino acids will help offer a steady supply of nutrients that are propelled by DNA-RNA to nourish and rebuild your tissues and cells. These amino acids are unique in that they can specifically offer protection against "cross-linkage," which is the very root-cause of much arthritis-like distress.

A CELLU-METRIC PLAN FOR ARTHRITIS RELIEF

Cellular rebuilding can be a delicious and healthy program. Here is a simple plan that is aimed at providing your body with the needed working materials to create adequate and *balanced* protein for cellular rebuilding and *protection* against malformed cross-links. It is a *Cellu-Metric Plan* because it rebuilds your cells, step-by-step, from the very beginning:

ANIMAL PROTEIN

Once a day, eat a concentrated protein food from any of this group: lean meat, fish, poultry, egg white. (Egg yolks may be eaten twice a week for cholesterol-counting.)

DAIRY PROTEIN

Daily, have several glasses of skim milk for good dairy protein. For better metabolism, have yogurt, buttermilk or any other fermented milk. The natural ferment works to improve intestinal action and digestive vigor so that amino acids of other foods can work more efficiently to rebuild cells and tissues.

WHOLE GRAINS

Plant proteins such as *millet* and *buckwheat* are also prime sources of amino acids that work in *harmony* with other amines to provide a *balance* of cellular rebuilding. Daily, eat whole grain breads and cereals. Eat millet and buckwheat in the form of cooked cereals to give your body the vital plant or grain proteins that appear to *complement* other proteins. This protects against a shortage that might otherwise predispose toward cross-linkage and arthritic symptoms.

BEANS, PEAS

These are not always complete, but they offer highly concentrated and high-quality proteins which are rendered "complete" when combined with other foods. Several times a week, have a plate of any type of cooked beans or peas together with your other protein foods.

BEVERAGES

Prepare fresh raw fruit and vegetable juices. These have some amino acids, but their great value is in their enzyme and vitamin content. Together with their minerals, they work upon consumed proteins to *digest* them and transform them into *usable* amino acids for cellular rebuilding. If you cannot make your own raw juices, then drink bottled or canned juices as available at health stores or at most supermarkets. Select those that have *no* salt, *no* sugar and *no* additives, since these ingredients destroy nutrients and inhibit the rebuilding powers of the amino acids. Other beverages may be herb tea, coffee substitutes such as Postum, or any other coffee substitute sold at almost all health stores or supermarkets. For any sweetening, use a bit of honey.

FRUITS, VEGETABLES

These are essential in your *Cellu-Metric Plan For Arthritis Relief.* Raw fruits and vegetables contain vitamins and enzymes which transform eaten protein into metabolized amino acids. Without these raw fruits and vegetables, ingested protein remains a "lump" or only partially metabolized and ineffective. Many arthritics do show a history of eating cooked foods with hardly any healthful raw foods. To protect against this problem, it is important to eat a raw fruit or vegetable salad with each and every meal. You may even enjoy a large raw salad sprinkled with Brewer's yeast and wheat germ as a self-contained good protein meal in itself, brimming with the needed vitamins and enzymes to transform the protein into cell-rebuilding amino acids. Some green leafy vegetables (the darker the green, the more concentrated the nutritional elements) are almost complete proteins of a very high *Cellu-Metric* order. They are a "must" in your program for protection against protein-starved "cross-linked" cell formation.

Benefits: This tasty *Cellu-Metric Plan for Arthritis Relief* helps stabilize cell membranes, and offers protection against cross-linkage by making available *balanced* amino acids for better cellular rebuilding. The *combination* of nutrients offers a balance so that proteins can be transformed into usable amino acids, which then work in *harmony* to rebuild the cells so that they are less vulnerable to becoming missynthesized, a problem occurring in many arthritic disorders. It is the natural way to help protect against arthritic unrest.

FIVE MIRACLE PROTEIN CELL FOODS

When your body has proper protein nourishment to guard against cross-linkage of cells, your limbs move with greater flexibility. There are five special groups of protein foods that appear to offer this benefit in cellular rebuilding. In your *Cellu-Metric Plan For Arthritis,* these five groups should be featured in your daily food program:

1. *Vitamin B-Complex.* Brewer's yeast, liver, wheat germ, bean and seed sprouts. *Benefits:* They help energize the transformation into more usable amino acids for better cell replication.

2. *Vitamin C.* Most citrus fruits and fruit juices, berries, bean and seed sprouts. *Benefits:* Works with protein to form collagen, the substance required for rebuilding normal cell membranes.

3. *Vitamin E.* Cold-pressed, non-processed fruit-vegetable-seed-nut oils. Also high in wheat germ oil, green leafy vegetables, egg, muscle meats, fish, whole wheat products. *Benefits:* They join with protein to help wash away cellular debris so membranes can be built without "wreckage" and with protection against cross-linkage.

4. *Sulfur Amino Acids.* This type of protein is found in egg yolk and egg white, muscle meats, cabbage. *Benefits:* They are compounds that help the body promote better molecular replication. They act *specifically* upon cell walls and tissues and are Nature's best protection against arthritic-caused cross-linkage of cells.

5. *Selenium Amino Acids.* Found in tuna fish, herring, Brewer's yeast, wheat germ, whole bran, broccoli, onion, cabbage, tomato. *Benefits:* This type of amino acid acts as a "resorter" or the substance that promotes and creates molecular wall replication. The cell organism needs this amino acid in order to guard against cross-linkage. In some situations, selenium amino acids may help break up incorrectly synthesized proteins and use new DNA-RNA compounds to build new and healthier membranes.

Easy Way to Use Five Miracle Protein Cell Foods for Your Cellu-Metric Plan for Arthritis: Each and every day, each of your three daily meals should contain one or more items from each of these five groups. Vary these items for better absorption and to offer your body a wide selection of different nutrients. Because eggs contain concentrated cholesterol, eat just two a week. But eat egg white often since it is cholesterol-free but protein-high! It's the tasty way to feed top-quality amino acids to your cells so they can repair smoothly, and minimize the risk of arthritis-connected cross-linkage!

HOW AMINO ACIDS HELPED CORRECT STIFF HANDS, ACHING WRISTS

Valerie G. was an active caterer. She would arrange all types of parties, celebrations, affairs from start to finish. She often boasted that she had so many assignments, she never had to cook a meal at home. It was true. She planned huge meals and menus for her customers, and would then eat these meals as the affair was being held. But there were problems that reacted with her health.

Limbs Feel Stiff, Aching. For weeks, Valerie G. felt that her hands were stiff. Her wrists ached. At times, she could scarcely bend her fingers to hold a pencil! She was frequently tormented with shock wave pains, knife-like, if she suddenly reached for an object or tried to grip it with her fingers. She would feel the same wrenching pain shooting right up through her arms and in her shoulders. She feared that this increasing stiffness might so hamper her work, she might become an unwilling retiree! Valerie G. felt the same painful spasms when she put her stiff hands into cool water. It was unnerving, to say the least. She tried exercises which offered temporary help. She maintained her diet was adequate since she always ate heartily. But the foods she ate did not adequately nourish her cells!

Cooked Foods a Clue to Cellular Misbehavior. At her catered affairs, much of the foods were processed, precooked, frozen, very much cooked and overcooked. At such affairs, fresh foods were difficult to come by. It was always "convenient" to have precooked foods ready. Valerie G. was living on *cooked* foods and this gave her protein, but she lacked the needed nutrients and the specific amino acids that would *break down* the protein into a form to offer nourishment to her cell walls. Literally, her cell membranes were "starving" and they replicated in "cross-link chains" that hampered free, flexible fingers, hands, wrists and joints. This was a clue to her cellular misbehavior.

Simple Corrective Program: Valerie G. made a surprisingly easy recovery by using this program: each and every meal began with a RAW fruit or vegetable salad. If none was provided by the catering hall, she bought some at the nearby market, washed it at home, and brought it with her for eating *before* and *after* a cooked meal. At home, she would drink fruit and vegetable juices throughout the day. Wherever possible, she would pass up any cooked food and prefer a raw fruit or vegetable salad with a chopped hard-boiled egg (no more than two or three eggs per week). Often, she would have a whole grain *breakfast* at home (she made time for this health restorative program) together with yogurt and fruit slices. *Lunch* would be a chopped chicken or turkey salad with assorted seeds and nuts. *Dinner,* if at home, would be fresh broiled fish, leafy green vegetables and a raw salad made with cold-pressed seed oils. As a *nightcap,* she would mix some Brewer's yeast in a vegetable juice, stir vigorously, then sip slowly.

Recovers Full Use of Arms, Hands, Wrists, Fingers. Valerie G. followed this amazingly simple but powerfully effective protein-ization program for several weeks. She had one basic rule: cut down on intake of cooked foods and eat as much raw foods as possible. As a result, the amino acids in these foods were able to be better metabolized, then used for better cellular replication. Her bone matrix and the millions of tissues and cells in the joints of her arms from the shoulders to her fingertips were able to be more *correctively* repaired, and cross-linkage cells were unlocked—and so were her arms and fingers. Now she had full flexibility of her hands and she felt good, and glad, all over—thanks to amino acid cellular replication and freedom from arthritic-related cross-linked cells!

HOW EVERYDAY PROTEIN FOODS REVERSED THE "ARTHRITIS-AGING CLOCK"

Burton McK. was in his early 60's, yet he looked and felt as if he had few years left to his life. Burton McK. was stooped over. He walked with a bent gait. He had to use a cane. Often, he hobbled along, with one hand on his bent over aching back, the other painful hand gripping a cane. He made a pitiful picture in his town. Everyone (including Burton McK.) said that it was arthritis and an accepted consequence of "growing old." Yet, Burton McK. knew of young folks who had arthritis and wondered if age was as important a factor as was erroneously believed. He set about to find ways to improve himself.

Refined Sugar, Refined Starches Are Villains. Looking around in his local luncheonette he saw dozens of folks of all ages, imbibing sugary pastries, refined carbohydrate foods, gulping down sugary cola drinks or endless amounts of sugary coffee and other such unhealthy beverages. Burton McK. could not point any arthritic accused finger at them. He, too, ate this way. Could faulty metabolism be traced to refined sugar-refined starch intake and could this be a cause of his arthritic distress? He could find out only one way. Change his eating program.

Sugar-Starch Free Program Helps Reverse His "Arthritis-Aging Clock." To help extend the life-span and *reverse* the factors that caused ticking of the "arthritis-aging clock," it was necessary to *eliminate* all refined sugar and refined starch from his program. This was the start. Next, Burton McK. followed this easy plan:

Every day he would eat as many raw fruits and vegetables as possible. He ate lean meats, fish, turkey, chicken, seeds, nuts and beans. Cold-pressed and non-processed vegetable oils with fruit juice were his salad dressings. He would drink fruit and vegetable juices (made at home or when purchased, sugar-free with no additives, no synthetic sweeteners which are chemicals and cellular-destructive). For a hot beverage, he would drink herb tea sweetened with honey. Or he would drink Postum or any coffee substitute. All of these items are available locally at just about any corner grocery or food market.

Straightens Up, Walks Erect, Discards Cane. Burton McK. found this adjustment was easy to make. He satisfied his "sweet tooth" with honey, some blackstrap molasses, maple syrup, and lots of luscious juicy good fruits. The addition of natural herbs and spices put new flavor into most of his foods. But most important, he discovered that this simple program soon eased his aches, his fingers became more flexible, he was able to walk straight, and he could throw away his cane, and walk like a young man. Indeed, he looked and felt better than men half his age who were already succumbing to premature cellular decay and cross-linkage arthritis. The easy program had helped reverse his "arthritis-aging clock" and given him a new lease on life.

Secret of Burton McK.'s Recovery. Refined sugars and starches cause a speedy rate of metabolism. They further *displace* protein and may even cause it to be stored and removed from the liver before the billions of body cells are able to derive needed amino acid nourishment. As a consequence, the cells are undernourished, and must draw upon their own reserves for rebuilding. This inferior or diluted type of protein causes cross-linkage. When much of the body is "repaired" in this manner, it is like being "patched" up. The body cannot perform as youthfully as if all cells were fully nourished and replicated. So when Burton McK. made the change and *eliminated* refined sugars and starches, his body could get at the protein and start bone-cell-tissue rejuvenation.

Specifically, the vitamins-enzymes-proteins in the healthful foods improved membrane stability and guarded against waste infiltration. Protein used the Vitamin C to improve the membrane stability, and also to produce more collagen to protect against the "choking off" reaction on the tissues and send nutrition to the cells so they could suitably repair *without* cross-links, which are the villains in arthritis!

Burton McK. found that life was worth living, thanks to this simple secret. *Avoid refined sugars and starches, and enjoy a balance of healthful vitamins-minerals-enzymes-proteins for cellular stability and youthful replication.* Give your cells the working materials for self-replication and they will give your body a smooth working youthful flexibility with forever young flexibility!

The root cause of arthritis distress is a systemic constitutional ailment of cellular disintegration and malformation of cross-linked (chained) membranes. These biological alterations can be halted and corrected by nourishing the billions of cells and tissues so they can give better flexibility and freedom from "blockage" caused by these chain-like cross-links. Proteins and their metabolized substances, amino acids, are the life's blood, so to speak, of the cells. Feed them protein and they can help build a youthful body that is free from stiffening.

Cellular nourishment is one important key secret in an extended youth cycle. Use but do *not* abuse your body. Eat wholesome foods. Keep yourself physically active. Any infections should be corrected.

Reduce and dispose of severe emotional and physical stress. Get adequate rest. Together with cellular rebuilding, it will help give your body the raw materials out of which lifelong health may well become the *possible dream.*

IMPORTANT POINTS

1. Arthritis is often a metabolic error traced to protein-starved cells that do not knit properly, and become cross-links that act as "chains" to inhibit full joint flexibility.

2. A protein balance creates more amino acids for DNA-RNA cell feeding substances. Eggs offer *three* basic amino acids that are specific for healthful cellular rebuilding and protect against arthritic distress.

3. Follow the simple *Cellu-Metric Plan For Arthritis Relief.* Everyday foods can become miracles of cell builders and protectors against arthritis-related cross-linkage.

4. Daily, eat one or more items from the special Five Miracle Protein Cell Foods for best protection against cross-linkage.

5. Valerie G. used simple protein foods and an easy-to-follow eating plan to help correct problems of stiff hands and aching wrists.

6. Burton McK. reversed his "arthritis-aging clock." He threw away his cane and walked straight and tall by avoiding *refined* sugars and starches and eating the foods listed in the *Cellu-Metric Plan.* Simple, yet youthfully invigorating. His body cells replicated *normally* as amino acids washed out cellular wastes and re-did cross-linkage, giving him a "new body."

8

WAKE UP A
TIRED BLOODSTREAM WITH
MIRACLE PROTEIN

You are as young as your bloodstream! This is the river of youthful life that gushes through and nourishes every part of your body. When the cells of your blood are adequately nourished with protein, they can help protect you against the aging process by transporting essential nutrients from one end of your body to the other.

HOW MIRACLE PROTEIN ENERGIZES YOUR
YOUTH-BUILDING BLOODSTREAM

Protein-nourished cells in your blood transport food from your intestinal tract and carry it to the most essential components of your body. These protein-nourished cells supply water and oxygen to your organs, transport the waste products produced by biological activity, and prepare them for elimination. The protein-nourished cells carry biological messages to all parts of your nervous system. The cells act as a medium for distributing nutrients throughout your entire body.

Your blood needs protein to help regulate the normal water content in your tissue fluids. The blood contains plasma proteins which are needed to maintain a healthful acid-base balance in your system.

Fat Metabolism. The blood uses protein to combine with insoluble fats to build complex molecular networks known as lipoproteins. This enables fats to be transported to their locations without pausing or delaying to form globules. Adequate protein in the blood is the natural way to protect against the buildup of excessive fatty deposits.

Better Breathing. The blood uses protein to transport oxygen from your lungs to your cells and then return carbon dioxide to your lungs for elimination and cleansing. This is made possible by a combining of iron plus protein to form *hemoglobin,* the red coloring matter of your bloodstream. This iron-plus-protein combination makes up to 17 percent of your bloodstream. While blood is distributed through the capillary network of lung tissue, this iron-plus-protein *hemoglobin* takes up oxygen breathed in, and sends it in a form that can be used for nourishment by the rest of your body cells. An adequate amount of iron with protein can create *hemoglobin* and improve breathing abilities.

Healthful Acid Balance. Protein is needed by the blood for vigor in carrying carbon dioxide away from the cells. Protein invigorated blood then uses some carbon dioxide to partially blend with the water content of the blood to form carbonic acid. The balance is then joined together with blood proteins. This process helps the bloodstream maintain a normal and healthy acid balance. Protein later helps the blood discharge excess carbon dioxide through the lungs or veins, so that the bloodstream can continue remaining a healthful acid base.

Nourishes Internal Organs. Protein-invigorated blood cells transport nourishing substances from the intestinal tract to most of the billions of cells throughout your body. They especially send these nutrients to your major and minor body organs. There is an internal network interdependence between your internal organs and the various systems, with the blood cells nourishing all of them. A protein deficiency means that the malnourished blood cells cannot properly feed most of your organs and inadequate health may result.

Washes Cells. Cleanses Body. Protein enables blood to carry waste products away from the body. In addition to "washing" the cells free of excess carbon dioxide, blood protein cleanses by helping to wash away such cellular by-products as urea, uric acid, phosphoric acid and lactic acid. Protein prompts the release and elimination of these wastes through the kidneys, urinary tract and other organs of elimination.

A well-nourished protein-ized bloodstream offers this built-in protection against illness. Feed your blood cells a balanced diet with emphasis upon protein, and they can help wake up a tired bloodstream and alert it to flow through your body with youthful alertness.

HOW PROTEIN HELPS YOUR BODY RESIST INFECTIONS

Your blood uses amino acids to form three basic plasma proteins within your liver—*albumin, globulin* and *fibrinogen*. These proteins join with gamma globulin (formed in the lymphoid tissue of the spleen) to form infection-fighting antibodies within your system. These special types of plasma proteins also help in the clotting of blood, and promote better healing and speedier recovery from infection. These plasma proteins also help neutralize bacterial poisons and detoxify such substances and build natural immunity.

Infection Fighting White Cells. White blood cells (leukocytes) need protein so they can destroy invading microorganisms and serve as a protection against infections. When your blood has sufficient protein, it can send *more* white blood cells to your lungs because this is the site where bacteria and other infectious substances may invade the system. A well nourished system with sufficient protein means the bloodstream can send protein-ized white blood cells in more ample profusion to those body portions where they stand as sentinels to guard against infections. These are some of the unique ways in which the bloodstream can be alerted so that it can offer healthful protection against the onslaught of age-causing factors.

Feed protein to your blood cells and your youthful bloodstream will help keep you young in body and in mind.

PROTEIN + IRON = YOUTHFUL BLOOD CELLS

Iron is a mineral that is needed by protein for blood nourishment. Iron combines with protein to form hemoglobin, the oxygen-carrying component of your red blood cells. Your body will hold some of this in storage, in the reticulo-endothelial cells of your liver and other vital body organs. A shortage or even a decrease of iron availability means that your blood cells may be unable to make a sufficient amount of youth-building blood cells. The key to youthful cells is in a simple combination: *protein + iron for youthful blood cells!*

OFFICIAL U.S. GOVERNMENT RECOMMENDATIONS
FOR BLOOD-BUILDING IRON

In the official publication, *Family Fare*[1], the U.S. Government tells us, "Iron is needed by the body in relatively small, but vital

[1]*Family Fare*, U.S. Department of Agriculture, Washington, D.C. Home and Garden Bulletin No. 1, May, 1970, page 11.

amounts. It combines with protein to make hemoglobin, the red substance of blood that carries oxygen from the lungs to body cells and removes carbon dioxide from the cells. Iron also helps the cells obtain energy from food."

Recommended Sources: "Only a few foods contain much iron. Liver is a particularly good source. Lean meats, heart, kidney, dry beans, dry peas, dark-green vegetables, dried fruit, egg yolk and molasses also count as good sources. Whole-grain bread and cereals contain smaller amounts of iron, but when eaten frequently become important sources."

In your cellular regeneration program, these iron-containing foods, as recommended by the U.S. Government, are most essential for good blood rejuvenation.

THE MIRACLE PRO-IRON DRINK THAT REJUVENATED THE BLOOD CELLS

Susan H. tried to live healthfully, according to the laws of Nature. But she was constantly catching colds. If someone just breathed heavily in her face, she soon came down with sniffles, a sore throat, runny nose, racking coughs, and a cold that just lingered and lingered. This was just one of her problems.

Susan H. was always feeling cold in her hands and fingers. Even in warm weather, she had to wear a sweater when doing her gardening in the sunshine! She always wore gloves, else she would become so chilly, she would start chattering.

Wounds and scratches did not heal rapidly. When she cut her hands while doing housework or the dishes, the bleeding took long to clot. Her manicurist refused to do her nails because one little cuticle scratch meant copious bleeding even when bandages were applied. Susan H. had pale cheeks and bluish lips, too. Fortunately, she recognized Nature's warning symptoms. She followed a surprisingly easy but amazingly effective cellular rebuilding plan. She prepared a special drink.

Miracle Pro-Iron Drink

In a glass of tomato juice, dissolve 2 tablespoons of desiccated liver. (The whole of liver but vacuum-dried so fat and connective tissues are removed. Available at most health stores.) Add a squeeze of lemon or lime juice for a piquant taste. Mix thoroughly. Take one *Miracle Pro-Iron Drink* in the morning. Take another one in the evening.

Benefits: The malnourished blood cells used the *combination* of protein and iron to build better red and white cells, the builders and protectors in the bloodstream. This easy beverage offered super-nourishment to Susan's iron stores and boosted protein absorption. Soon, she could resist colds. She had tingling warm hands and fingers, and gave up wearing sweaters or gloves. Wounds healed so rapidly, she was welcomed back by her manicurist. Tiny scratches were hardly noticeable. Her cheeks perked up with the color of Springtime roses. She felt and looked radiantly youthful, thanks to the *Miracle Pro-Iron Drink* that sent a concentrated supply of nutrients needed for cellular replenishment.

THE SECRET "GARDEN GREEN DRINK" THAT REJUVENATES THE BLOODSTREAM

Arlene P. felt "half alive." She walked with a stooped gait. Her skin became prematurely wrinkled. There were noticeable creases and dark blotches beneath her eyes, giving them a "baggy" look of aging. Her skin color was waxy and pale. She always had beads of perspiration on her cold-looking skin. To touch her hand brought a response from the other person, who would remark that her hand felt "as cold as ice." Arlene P. was always huddling near her radiator for warmth. Nightly, she needed an electric blanket. Her condition worsened. She was told that an adequate protein program, combined with the other needed nutrients, would help her perk up. But she needed to have a combination of protein plus iron so that the red and white cells in her bloodstream could become supercharged, increased in number, and send a throbbing feeling of warmth and vitality shooting through her body. She set about to follow a simple two step program:

1. **Iron + Protein Breakfast.** A whole grain cereal such as oatmeal, whole bran, granola (sold at most supermarkets and health stores) in skim milk, sprinkled with unflavored gelatin powder. She finished with yogurt and fruit slices. *Benefits:* This gave her good grain and animal protein that became better metabolized by the enzymes in the yogurt and fruit. The iron of the whole grains could now join with protein, replenish the depleted stores in the bloodstream and "fortify" the body with needed cells for overall invigoration.

2. **Secret Garden Green Drink.** In a juice extractor (or a blender) liquefy a variety of dark-green leafy vegetables. The darker the

vegetables, the greener the juice will be and, subsequently, the more potent the supply of iron + protein. Daily, Arlene P. would drink at least three glasses of this *Secret Garden Green Drink.* She was soon able to respond with renewed vitality. Her skin became smooth. Blotches and creases were no longer noticeable. The nervous beads of perspiration were gone. Her hands were warm to others and to herself. She walked with a straight figure. Arlene P. was so protein-ironized, her blood cells felt like "singing" their radiant warmth and youthfulness. Life now was joyful. *Benefits:* Plant iron and plant protein in the drink joined together to send a stream of metabolized amino acids to the blood cells so that they could become replicated, repaired and regenerated. Once the iron-protein combination was metabolized, her body was rewarded with living cells that made her radiate warm health. The deficiency was corrected with the nutrients in the *Secret Garden Green Drink.* Just a few plant ingredients, but chock full of brimming iron and protein that become food for the starved blood cells.

THE FRUIT THAT WAKES UP AND POWERHOUSES YOUR CELLS

Gordon L. would tire after just some routine tasks. As a school-teacher, he had a full schedule and had to continue working over his students' exams late at night. He was often on call for lectures. He did some travelling on weekends to take post-graduate courses for needed credits for eventual promotion. But Gordon L. became inefficient. He tired easily. His brain felt fuzzy. His thoughts were vague. He made errors. His students giggled over his fumbling. He was called in "on the carpet" by his supervisor and told to "shape up" or they would have to release him. Gordon L. did not want so many years of preparation to go down the drain. He had tried everything. Since he was teaching basic biology to his students, he knew that his bloodstream could have been malnourished. He also knew of one fruit, *the apricot,* that was the "star" blood cell food of almost all known fruits. So he prepared a delicious sunshine-colored tonic as cell food.

Golden Blood Building Tonic: At any health store or organic food outlet, purchase sun-dried golden apricot slices. They *must* be sun-dried, free of sulphur dioxide or hydrogen peroxide. These two chemical preservatives are destructive to body cells and tissues. So purchase sun-dried apricot slices. The night before using them, put them in a glass bowl filled with recently boiled water. Cover. Let remain overnight.

Next morning, drink this *Golden Blood Building Tonic.* Almost immediately, you will feel a tremendous surge of vitality pouring through your bloodstream. If you drink this *before* any breakfast, you'll be doubly rewarded since the nutrients can work without interference in building and rebuilding body cells. Also, eat the fruit slices in a whole grain cereal for breakfast.

Benefits: Just 1 cup of these sun-dried apricots and the juice will give you 8 grams of plant protein together with 8.2 milligrams of iron. You are also given 16,350 units of Vitamin A and 19 milligrams of Vitamin C. These units *converge, combine* and *blend* with natural enzymes so that they can become *built* into your blood cells and tissues. Of all fruits, apricots appear to offer the maximum benefits in this regard.

Gordon L. found that a daily *Golden Blood Building Tonic,* together with other good protein and iron foods, performed a speedy regeneration. These nutrients rebuilt and regenerated his "tired" blood cells. A well-nourished and protein-ized blood system worked with the other vitamins and minerals to promote youthful energy. His thoughts were alert. He no longer fumbled. He was so alert and vigorous, his students complained he was overworking them! Soon, he received not one but three promotions. He now feels "younger than young" and is the envy of his co-teachers and students, too. His *Golden Blood Building Tonic* proved to be a goldmine of healthy success!

THE PROTEIN WAY TO NATURAL BLOOD WASHING

Symptoms of aging are frequently traced to a loss or decline of blood cellular reserves. This type of cellular aging begins very early in life, and if corrected in time through a unique and natural way of "blood washing," it can offer hope for a limitless youth!

Free Radicals: Villains in Bloodstream. A free radical or "root" is a fragment of a molecule that has been broken off from its source during a biological process. It is comparable to looking upon litter and debris in a lake, stream or river, or in your own kitchen sink! The natural process of oxidation in the bloodstream causes this type of breakage and sends endless amounts of these roots and debris throughout the body's river of life.

In effect, excessive radicals in the bloodstream can be likened unto internal pollution. Or, the bloodstream becomes polluted!

These bloodstream pollutants cause aging by interfering with the actions of DNA-RNA in replicating cells and tissues. They block the normal repair and may lead to problems of cross-linkage or mis-synthesized protein repair of the cells. The more internal pollutants allowed to remain in the bloodstream, the more damaged the skin cells remain and the aging process increases in insidious intensity.

Therefore, cellular aging may be traced to alterations of DNA-RNA by the free radicals or internal pollutants. These unhealthy fragments cause proteins to be incorrectly synthesized, and this may lead to the death of the cells. Unless unchecked, the increase of these pollutants can actually destroy the body through death of the billions of cells throughout the bloodstream. *It is a slow and gradual decline that is noted during the aging process. Bit by bit the bloodstream becomes polluted. Little by little the life processes lose youthful efficiency. Drop by drop the ebb of life recedes.* But Nature has made protein available for cellular repair. Protein, however, needs another substance to create this natural blood washing. The other substance is Vitamin E in the form of any cold-pressed non-processed seed oil.

Protein + Vitamin E = Blood Washing. Polluted cells may often lose their ability to decode the instructions of DNA-RNA to properly become healed. This suggests that protein needs Vitamin E for "instructions" or "blueprints" to perform a unique action: Protein uses the alpha tocopherol of Vitamin E to create an anti-oxidation act to reduce the amount of pollutants formed by oxidation. Protein + Vitamin E work to nullify the *lipofuscin* (age pigment formed by the solution of pigment in fat) and then help to wash out the debris from the bloodstream. This combination has a unique effect in helping to guard against accumulation of such debris.

Secret Power: Protein uses the substances in Vitamin E to delay unnatural oxidation of the polyunsaturated fatty acids which can otherwise cause solidification of tissue proteins and more breakage. This helps wash out the excessive "sludge" from your rivers of life, guard against blood pollution, *the natural way,* and helps keep your bloodstream as youthfully alive as your very body and mind!

Simple Blood Washing Program: For cooking and for salads, and with any protein food, try to use polyunsaturated oils. Wheat germ oil, peanut oil, safflower seed oil, corn oil, cottonseed oil and avocado oil are just a few of the high protein-Vitamin E oils sold in most health stores and just about any supermarket or food outlet. Use these plant oils in a large green leafy salad. Eat the salad when eating any protein meal. Combining proteins with the Vitamin E appears to allow them to *complement* each other. This exerts a beneficial blood washing effect by helping to wash away the decomposed, decayed and dead fragments of cellular byproducts.

Grain Foods Offer Natural Combination. Whole grain foods made from wheat, oats, bran, corn, beans, nuts and seeds are prime sources of the *combination* of protein plus Vitamin E. In this *natural blend,* the elements work to promote a cleansing action upon the bloodstream. It is healthful to eat whole grain foods as bread products, use whole wheat germ, and enjoy delicious whole grain breakfast cereals with milk or fruit juice as a healthful combination of protein with vitamins, minerals and enzymes. They work in harmony to create a scrubbing action upon pollutants in the bloodstream and help wash them out of your system.

Your "tired" bloodstream may be Nature's warning signal that this river has pollution and is in need of natural cleansing through a combination of protein with other essential nutrients. Wake up your tired bloodstream with protein washing. Let your rivers flow with a youthful sparkling vibrancy. You will radiate youthful good health from top to toe with healthy cells.

SUMMARY

1. Miracle protein nourishes your blood cells. Your bloodstream can then perform better fat metabolism, better breathing, maintain a more healthful acid balance, nourish your internal organs, wash your cells and cleanse your body.

2. Three amino acids are needed to nourish your blood to help your body resist infections and neutralize poisons.

3. Protein + Iron = Youthful Blood Cells. Feed your cells some of the government-recommended iron foods for better protein metabolism.

4. Susan H. was able to correct her cold-catching and cold hands by trying a simple *Miracle Pro-Iron Drink.* Just 3 ingredients, available locally, work a miracle of cellular rejuvenation.

5. Arlene F. felt "half alive" until she energized her body with a two-step program. She had an Iron + Protein Breakfast. Then she had a *Secret Garden Green Drink.* Rebuilt blood cells made her throb with youthful vitality and smoother skin.

6. Gordon L. used a simple fruit, available anywhere, as a "blood food" *or a Golden Blood Building Tonic,* that replenished his cells and made him younger than springtime!

7. Use protein with Vitamin E to wash the pollutants or "free radical" debris out of your bloodstream. It's the natural way to guard against cellular aging.

9

HOW RAW FOOD PROTEINS CAN HELP CONTROL CHOLESTEROL

Body cells and tissues must constantly protect against an *enemy* in their midst at all times. That enemy is cholesterol. In normal quantities that the body can assimilate, it is an important element. But when there is an excess of cholesterol, it can overrun the body cells and coat them, choke them and cause death. Cholesterol can become a serious enemy if it overpowers the body cells and tissues. For cellular health, cholesterol should be kept to a minimum.

What Is Cholesterol? It is a tasteless, odorless, white fatty alcohol found in all animal fats. It is a member of the chemical family of fatty types of alcohols generally known as sterols. In pure form it is a white wax-like material. It does not dissolve in water.

The word cholesterol orginates from *chole,* Greek for "bile," and *sterol,* from the Greek word for "solid." That is just what cholesterol is—a fatty "clump" found largely in the digestive bile. (It is the stuff of which some gallstones are composed.) It circulates throughout the bloodstream and if piled up in excess, tends to to coat the cells, tissues and arteries of the body, hindering free flowing metabolism.

Risk to Cells of the Heart. In a *healthy* heart, the cells and tissues lining the coronary arteries are clean and smooth. In an *unhealthy* heart, the cells and tissues of the arterial walls are clogged with cholesterol. The cells become thickened and rough. This thickening is caused, basically, by the slow deposit of the fatty cholesterol wax called cholesterol. Like the rust in a pipe, cholesterol clogs up the cells and tissues and tends to restrict the blood flow to the heart tissues.

Atherosclerosis. This word is taken from the Greek word *athere,* and means "mush" or "porridge". This describes the accumulated cholesterol in and on the cells and tissues lining the body's arteries. When these cells are "choked" or "clogged," then there is a buildup of fatty material so that protein cannot get through. The artery becomes less elastic. Problems compound. The "choked" arteries are unable to send through enough nourishment to the heart. There is a sudden call for more oxygen. But this is difficult because of the choked cells and tissues. Narrowed arteries and atherosclerosis occurs and the heart muscle is in danger. So is the very life of the individual!

RAW FOODS THAT HELP PROTECT AGAINST CELL-CHOKING CHOLESTEROL

A *key* to protecting your body against such a cell-choking *over*-abundance of cholesterol is to restrict excessive amounts of animal foods. At the same time, substitute or replace with plant or raw foods that require *no* cooking so that you will have *prime protein* that can become metabolized and then join with essential fatty acids needed to inhibit excessive cholesterol buildup.

Some helpful cell-tissue nourishing foods that work to " melt" or control cholesterol buildup are these:

Lecithin. This is a bland, water-soluble granular powder made from de-fatted soy beans. Biochemists call it a *phosphatide*. This means that it is an essential component of all living cells and tissues. Lecithin is a *raw protein food* (sold at most health stores), a natural substance found in your brain, nerve tissues, cells of the endocrine glands, kidney cells and heart cells. Lecithin helps guard against cholesterol by performing these actions:

1. *Lowers Surface Tension.* Lecithin protein lowers the surface tension of aqueous solutions. It acts as an emulsifying agent, capable of dissolving cholesterol deposits on and within cells and tissues.

2. *Improves Absorption.* Lecithin protein increases the digestibility and absorption of fats because of its emulsifying abilities. The protein enhances the absorption and utilization of Vitamin A, which is built into your cells and tissues and protects against cholesterol.

3. *Metabolizes Fats.* Lecithin protein boosts fat metabolism. Its enzyme, lecithinase, sets free choline, which works with protein to help protect against excessive fat accumulation in the liver. Lecithin protein is needed to invigorate choline, and to keep cells and tissues clean.

4. Better Liver Function. Lecithin protein protects against deranged fat metabolism, helps feed the liver cells a balanced amount of nutrition and protects against cholesterol buildup.

5. Miracle Cleansing. Lecithin protein attacks cholesterol by using its own lipotropic substances to clear up vitreous opacities (the forerunner of cholesterol deposits—a colorless thickening), and cleanses by a " miracle" action so that the sparkling clean cells and tissues reward the body with resilient and free-flowing arteries.

How to Use Lecithin: This raw food can be sprinkled over whole grain cereals for breakfast. Add whole wheat germ, as well. Or mix with a fresh raw vegetable or fruit juice, stir vigorously and drink as a healthful "fat washing" protein tonic. Mix lecithin granules when making a large raw salad. Add a bit of vegetable oil with lemon juice for a delicious meal in itself. Health stores also sell lecithin as food supplements. It is a natural food and should be used as part of your cell-washing program against cholesterol buildup.

THE CHOLESTEROL-FIGHTING PROTEIN FOOD FROM THE OCEANS

The Food and Agricultural Organization of the United Nations, as reported in the *U.S. Fisheries Marketing Bulletin,*[1] cheers the use of fish oils as a means of fighting cholesterol. Fish oils of any sort contain protein plus unsaturated fats which work in *harmony* to wash the sludge from the billions of body cells. The FAO of the United Nations tells us:

Most Effective Agent. "The unsaturated fats in fish have been demonstrated to be among the most effective agents known to man in reducing the elevated blood cholesterol levels commonly associated with heart diseases. Fish are the only source of animal protein food in which this type of fat is found in abundance."

Easily Used by Body. "Fish oils (protein) are uniquely well fortified with more of the unsaturated fatty acids than are vegetable fats or land-animal fats. Fish fat is easily digested and readily used by the body tissues."

Most Potent Source. "The polyunsaturated portion of most fish oils (protein) is much more potent than that of the vegetable oils. Some species of fish contain relatively high quantities of fish oil in the edible flesh so that the consumption of the fish results in the

---

[1] *U.S. Fisheries Marketing Bulletin,* Bureau of Commercial Fisheries, U.S. Department of the Interior, Washington, D.C., 1971.

ingestion of considerable fish oil. For example, some salmon average 15 percent oil. A 160-gram serving would furnish 24 grams (nearly one ounce) of salmon oil rich in polyunsaturated acids. This *neutralizes* any harmful effects of other saturated fatty acids that might be consumed at the same meal, and probably contributes somewhat to reduction in blood cholesterol level."

Protective Function. "Thus, use of fatty fish as a main course permits consumption of such foods as dairy and meat products, reducing the possibility of any harmful effects resulting from the lack of unsaturation in their fat."

Suggestions: Fish oil, sold at almost all health stores and corner markets, should replace the use of *hard* fats. Fish oil is a raw protein food that joins with its amount of unsaturated portions to act as a cell washing agent. Use fish oil mixed in a glass of vegetable juice. Or use a spoon or two on a raw vegetable salad, and sprinkle with lecithin for an all-powerful high-protein, fat-melting, cell-washing reaction.

Simple Program Helps Earl W.T. Guard Against Hardening of the Arteries. Earl W.T. felt short-breath. He was forgetful and absent-minded. Sometimes, he had to read a sentence several times in order to understand its meaning. Frequently, family members and neigh-bors had to repeat the same question to him several times because he could not grasp its meaning. It was believed that cholesterol-choked cells had caused a build-up of atherosclerosis and this choked off needed oxygen for his brain cells. Protein could not be transported through the choked arteries and he showed some symptoms of so-called senility. His understanding wife decided to do something other than criticize a problem that was out of his control. But it was within *her* control. She switched until Earl W.T. had a frequent raw food day with emphasis upon fish oils. *No hard fats* in the form of meat were permitted.

Program: Earl W.T. had a *breakfast* of whole grain cereals, sprinkled with lecithin, in skim milk with fruit slices. For *lunch* he had a large raw vegetable salad with fish oil and lemon juice for a dressing, sprinkled with chopped nuts and seeds. Grated applesauce with lecithin was a tasty dessert. For *dinner,* he would have a large fruit salad with more fish oil and lemon juice dressing. He could have as much yogurt (made from skim milk) with wheat germ as desired throughout the day.

Results: The high raw food protein content of these foods

"overcame" the encroaching cholesterol buildup, freeing the choked cells and tissues so that they could send needed blood and oxygen through the rejuvenated clean arteries to the brain and other components.

Earl W.T., on a reduced animal fat program, was soon so mentally alert, with such a vivid memory, he surprised his younger children with such trigger-quick reflexes of mind and body!

A raw food program should be built into the cell-tissue rejuvenation program. Aim for one day a week devoted to raw foods. It is a tasty way to offer your cells and tissues unusual *plant* proteins that perform a scrubbing action, so that arteries and cells are clean and healthy. Fish oils should replace animal fats and oils as much as possible. Take advantage of the unusual protein found in fish oils. They help wash away cholesterol with the vigorous force of an ocean tide!

HOW RAW FRUIT CAN CONTROL CHOLESTEROL

Fresh raw fruit, when eaten together with a protein food such as yogurt, cheeses made from skim milk, and assorted seeds and nuts becomes a *raw food meal* which offers a unique cholesterol-fighting action.

Protein will take the Vitamin C out of the fruit, during the process of metabolism, and use it to flush cholesterol from the inner cell walls of the arteries. It is the *equilibrium* of raw food protein with raw fruit Vitamin C that helps to decrease the excessive buildup of lipoprotein lipose. This, in turn, controls the high-serum triglycerides and atherosclerosis formation. Protein plus Vitamin C, when both are taken from a raw source, are as powerful as Nature created them, and they help prevent plaques from forming on the cells and tissues. Protein plus Vitamin C can seep through the plaques, and then is taken up by the cells for "washing" from within. Soon, the plaques can be broken down and washed out of the system—thanks to the steady action of plant protein with Vitamin C.

Suggestions: Several times a day, combine any seasonal raw fruit (citrus fruits such as oranges, grapefruits, tangerines, lemons, limes are highest in raw Vitamin C) with any raw protein food such as skim milk or yogurt. Or have a plate of seeds and nuts (high in protein) with fresh fruit slices. You will be giving your billions of cells and tissues the *combination* of substances needed to protect against and reduce excessive cholesterol.

FROM "SENILITY" TO "SECOND YOUTH" THANKS TO PROTEIN

Jennifer Y. was a pitiful picture. She was considered "senile." This is the saddest aspect of cellular aging when life continues but the brain is malnourished and cannot function. Jennifer Y. was said to be resigned to her fate of living out her best later years in a vegetable-like existence that offered little meaning to her, or others around her. She had difficulty in remembering telephone numbers, street addresses, appointments, dates, names and even faces of her loved ones. She sat for hours in a chair, staring at nothing!

Kindly Niece Creates Miracle Protein Plan. A kindly niece, Sara Y., noted that Jennifer Y. subsisted on prepared meats, heavy fatty foods, and processed foods. Since Jennifer lived alone and had no one to cook for, she did not want to "bother" preparing meals just for one person, herself. So years and years of eating heavy, fatty foods, with very few vitamins, minerals or good protein, had taken their toll. The fat buildup "choked" the cells of her arteries, and oxygen deprivation made her brain "starved" and she displayed symptoms of "senility."

Fresh Raw Fruits, Vegetables, Seeds, Nuts Daily. Sara took her aunt off all processed and heavy foods. She gave her daily meals consisting solely of *raw* seasonal fruits, vegetables, seeds and nuts. Fish oils were used for salad dressings. Fruit and vegetable juices were her beverages. Sara mixed large amounts of lecithin with powdered seeds and nuts and sprinkled it into a juice for Jennifer to drink.

Boosts Protein Assimilation with Fruit. Sara insisted that Aunt Jennifer eat yogurt three times daily. Each cup of plain yogurt had fresh fruit slices. This gave her animal protein with plant vitamins that worked together to create the cellular washing action. This *combination* acted synergistically; that is, produced an increase in the action of the metabolism of the protein into amino acids while energizing its use of Vitamin C. A simple, yet effective health-boosting function.

Recovers Memory. Becomes Youthfully Cheerful. Jennifer Y. soon responded with a more vivid memory. She remembered everything. Her wit was sharp and pleasant. She was so youthful that she joined several clubs as outlets for her new-found energies (mental and physical.) Once her billions of cells and tissues used protein and other elements to cleanse away the choking cholesterol, her arteries became flexible roadways through which brain-feeding protein could be carried. She still restricts animal proteins, and has two days a week devoted to a raw food protein program as outlined above. She says that protein helped her go from "senility" to "second youth."

A CELL-TISSUE FEEDING PROGRAM FOR FOLKS OVER 55

Over 55? You can enjoy the best years of your life with a good cell-tissue feeding program. A. M. Nelson, M.D., treated 206 folks with coronary artery ailments and discovered that a fat-controlled diet *high in protein and in seafoods* (except shellfish) can be of enormous benefit for folks over 55. Writing in *Geriatrics*,[2] Dr. Nelson offered this 5-step plan. (A meal plan for any single day includes selections from these food categories. It enables you to have food variety while you are giving your cells and tissues a balance of needed nutrients. Not all foods are raw, but they serve as a balance for better health.)

1. *Dairy Products.* Milk or milk products. One pint or more of non-fat (skim) milk or buttermilk is recommended. Low-fat cottage cheese and certain low-fat cheeses are also allowed. Read the label to see if it is made from skim milk, which makes it acceptable.

2. *Animal Protein.* A total daily allowance of 6 ounces (about one cup) of cooked lean beef, lamb, veal, chicken or turkey is permitted. The patient is encouraged to substitute seafood for meat, *in unlimited amounts,* at least three times a week.

3. *Plant Foods.* The patient should have dark green or yellow vegetables, tomatoes or citrus fruits at least four times a day.

4. *Grain Proteins.* The patient should eat four or more servings of whole grain products daily.

5. *Unsaturated Fats.* A daily intake of three tablespoons of oil, including that used for cooking, is the amount permitted in this diet. Safflower and corn oils, corn oil and safflower mayonnaise, or safflower and corn oil margarine may be used.

This simple 5-step program uses basically raw foods, and acts to send a supercharging of proteins and other nutrients into your body to protect against cholesterol buildup and keep your arteries (and you, too) looking and feeling zestfully young.

Since cholesterol is found ONLY in animal foods, you can readily

---

[2]*Geriatrics,* December, 1972, page 103, "High-Protein, Low Fat Diet."

appreciate the rewards of enjoying meatless protein from plant sources as a means for helping to correct or control cholesterol caused atherosclerosis and cellular blockage. Meatless protein foods help your body use up the store of cholesterol and keep levels to a suitable low. Animal protein in the form of meat and eggs may be used *infrequently* until the body has been able to cope with the cholesterol. Afterwards, enjoy meats and two eggs a week, but keep them to a minimum.

QUIT SMOKING FOR BETTER CELL HEALTH

When you smoke, you send toxic wastes through your lungs, directly into your bloodstream. These wastes attack the arterial cells and tissues. Tobacco smoke causes formation of debris or cellular litter such as free radicals. Furthermore, inhaled nicotine forces your body to release more adrenalin, which causes your fat cells to release more and excess "fatty substances" into your bloodstream. Cigarette smoking destroys cells and tissues and causes premature aging. Give up smoking for better cell-tissue health and lower cholesterol levels!

KEEP YOURSELF PHYSICALLY ACTIVE

Your cells have to be "exercised." You can do this by keeping yourself physically active. Regular walking, doctor-approved jogging, hiking or sports (choose relaxed sports such as golf, swimming, bowling) help protein metabolize cholesterol. Protein is invigorated by a body that is kept active, within reasonable doctor-approved limits. So do regular walking outdoors in good weather. It's the healthy way to alert your metabolism to work protein into amino acids, which then synthesize cell walls and create better arterial health.

Cholesterol is a natural body substance, but too much can build up fatty deposits in the arterial cells and tissues, making it difficult for the blood to flow properly. When this happens, "choked cells" become narrow and cannot transport nutrients to the heart and brain muscles and other organs. It is a major predisposing factor to coronary heart distress. Nature has a plan for controlling cholesterol. Protein and other nutrients can work in *harmony* to protect against cellular choking and control cholesterol.

IN REVIEW

1. Raw foods are *prime protein* sources of elements that can inhibit excessive cholesterol buildup.

2. Lecithin is a soybean food that offers raw protein to help control cell-cholesterol-choking in five different ways. Tasty way to protein-ize and protect your cell walls.

3. The Food And Agricultural Organization of the United Nations recommends fish oils as a high-protein ocean food that fights cholesterol and protects cells and tissues from atherosclerosis.

4. Earl W. T. followed an easy program to wash his cells. His breath became better. His memory improved. He protected himself with protein, and avoided hardening of the arteries.

5. Raw fruit with a raw protein food can help control cholesterol and rebuild body cells. Jennifer Y. used this program and went from "senility" to "second youth" in a very short time.

6. Folks over 55 should heed the 5-step cell-tissue program outlined by a doctor.

7. Quit smoking, keep yourself active, and give protein a chance to give you "forever young" arteries and freedom from excessive cholesterol.

10

NATURAL AND YOUTHFUL IMMUNITY AGAINST INFECTIONS WITH PROTEIN

Your body cells and tissues need a ready and waiting supply of amino acids in order to help give you natural immunity against such infections as an allergic attack, respiratory distress, bronchial unrest and colds and sniffles. The secret here is that an infection is generally a symptom of a virus or microbe attack on the cells and tissues. Metabolized protein in the form of amino acids works with available vitamins to create natural antibodies to fight the virus infection and inhibit its encroachment. Amino acids help create the formation of antibodies in the body and bloodstream, resulting in a fortress against virus attacks that cause infectious reactions.

THREE VITA-PRO ALL-NATURAL "COLD REMEDIES" FOR NATURAL IMMUNITY

To help build natural immunity and protection against winter or respiratory infections, your body needs a balanced food program, adequate rest and emphasis upon three unique "vita-pro," or vitamin-protein, *combinations* that can help stimulate production of antibodies.

Form Antibodies. These three all-natural "cold remedies" help form antibodies by stimulating the body's own defensive mechanisms.

(An antibody is a protein or gamma globulin, formed in the lymph or straw-colored portion of the bloodstream. It is needed for fighting infection. Having ample antibodies in the bloodstream is your fortress against infections.)

These three vita-pro natural remedies boost body production of antibodies. These then act as the first line of defense against infectious invaders in the bloodstream.

114

Secret of Vita-Pro Effectiveness. Antibodies can offer you natural immunity if they are (1) constantly available in our bloodstream, ready to fight foreign invaders such as virus germs, and (2) active and vigorous enough to move with vim and vigor, swirling throughout your bloodstream on their mission to "seek and destroy" foreign bacterial infections.

The following three vita-pro natural remedies are effective because they work in *combination*. They help your body manufacture needed antibodies. They give them the vigorous thrust so they can go rushing throughout your bloodstream on their protective mission. Build them into your daily food program to help give you year-round antibody protection against virus infections.

1. PYRIDOXINE + PROTEIN = NATURAL ANTIBODIES

Pyridoxine or Vitamin B6 works together with protein in the biosynthesis of nucleic acids, both DNA and RNA. The combination of pyridoxine *plus* protein sends sufficient nucleic acids to your billions of body cells and tissues. These "cell feeders" help produce the antibodies and other forces that can fight infectious wastes. Pyridoxine becomes better assimilated through the action of protein. As a result, the bronchial or respiratory tissues are then able to rebuild better, offer stronger protection against foreign invaders, and help send a healthy flow of antibodies shooting through the bloodstream to attack and destroy, and then wash out, the infectious germs. It's a natural way to build immunity.

Vita-Pro Bronchial Healer: Combine Brewer's yeast with one tablespoon of unflavored gelatin powder and mix together in a glass of tomato or any vegetable juice. Add a bit of lemon or orange juice for a taste boosting. Drink one glass in the morning. Drink another glass at noontime. A third glass just before retiring is beneficial, since many bronchial attacks occur during nighttime.

Benefits: The Brewer's yeast is a rich source of pyridoxine, which joins with the nearly complete amino acid pattern of the gelatin powder. Together, they are better metabolized by vitamins and minerals in the vegetable juice. This *combination* goes speedily into the bloodstream for cellular rebuilding and profuse antibody production. It is a natural way to rebuild the *bronchi* or air tubes of the lung. Bronchial arteries are made of cells and tissues and should be strongly nourished with the *Vita-Pro Bronchial Healer*. This also helps boost antibody production, and offers good basic immunity-fighting substances for use against bronchial disorders.

Pyridoxine Sources: Into your food program, add Brewer's yeast, organ meats such as kidney and liver, lots of green leafy vegetables, wheat germ, nuts and seeds. *Combine* these with good protein foods and you'll have a double-barreled natural fortress against virus attacks.

2. FOLIC ACID + PROTEIN = CELLULAR IMMUNITY

Folic acid is part of the B-complex family. It joins with protein to go to the millions of cells within the bone marrow. Here, both of these nutrients work to form red blood cells. These are then propelled by folic acid and protein to transport antibodies and provide a form of cellular immunity. Both of them are needed to help boost red blood cell health. The first line of defense is in a healthy supply of strong red blood cells. They can create *natural immunity*. A combination of folic acid plus protein can help give your body the needed ingredients to promote this cellular immunity.

Vita-Pro Cold Fighter: Prepare a glass of fresh *green* vegetable juice. (Use a juicer to squeeze your own, from cabbage, lettuce, celery. Or get it from a health store or supermarket, already canned or bottled.) Mix in some wheat germ, add desiccated liver and a sprinkle of lemon juice. Drink several glasses throughout the day, especially when you feel the approach of a cold.
Benefits: The green vegetable juice is a powerhouse of folic acid that joins with the protein of the wheat germ and desiccated liver to form a *complete* amino acid pattern. This *combination* is sent speedily to the bloodstream, which will use the combined vitamins and amino acids for rebuilding of the red blood cells. They can then act to nullify and weaken virus infections. They act to strengthen the permeability of the cell walls, which enables them to resist bacterial onslaught. The *Vita-Pro Cold Fighter* contains the needed blend of natural protein and vitamins that can help give you resistance against infections with the vigor of a youngster.

Folic Acid Sources: Your food program should include wheat germ, all green vegetables (these are prime sources), Brewer's yeast, liver and most organ meats. *Combine* them with protein foods and your blood cells can become super-nourished to give you cellular immunity, the key to protection against infections.

3. PANTOTHENIC ACID + PROTEIN = VIRUS PROTECTION

Pantothenic acid is an important member of the B-complex vitamin family. It is important in carbohydrate metabolism. Often, an excessive intake of *refined* sugars and starches (these are carbohydrate non-foods) displaces protein metabolism so that amino acids cannot create sufficient antibodies to protect against virus infections. Pantothenic acid is important for metabolizing carbohydrates, and then works upon protein, which is used for energizing the system and helping to create strong antibody fortification. A wise cold-fighting course is to *restrict* intake of *refined* carbohydrates. Get your carbohydrates from fresh fruits and vegetables and whole grains in a *natural* form that the body can accommodate. A ratio of *increased protein* and *reduced carbohydrates* can give your body a balance so that it can place better attention to metabolism of protein, and cannot be diverted by carbohydrate interference.

Vita-Pro Allergy Soother. Dissolve two tablespoons of Brewer's yeast in a glass of fruit juice. Add some wheat germ. Stir vigorously. Drink several glasses throughout the day. It's helpful to drink this *before* any allergic attack as a protective bulwark against allergens.

Benefits: The pantothenic acid in the Brewer's yeast works with the vitamins in the fruit juice and joins with the wheat protein. This creates a form of complete blending so that the nutrients can become better assimilated within the bloodstream. These vitamins help the body strengthen its cells and walls to deactivate the infecting microbes. The combination of pantothenic acid plus protein helps your cells strengthen by the formation of allergens, which are protein-like barriers that resist, and even weaken and eventually cast out microbe invaders. In this way, the *Vita-Pro Allergy Soother* sends protein to deactivate the microbe and strengthen the millions of cell walls and membranes to resist further attack. It is a healthful hope for *natural immunity,* thanks to the combination of vitamins and protein.

Pantothenic Acid Sources: Your daily food program should include lean liver, poultry (turkey is a good source), Brewer's yeast, whole grain breads and cereals, dark colored vegetables, salmon, eggs, mushrooms, peanuts, broccoli, soybean flour and all soybean products. Throughout the month, select foods from this group and combine them with healthful protein foods (many of the preceding contain this Nature-built combination), and help create natural immunity.

THE SIMPLE FOOD PROGRAM THAT GAVE BETTY J. NATURAL IMMUNITY

Living in the cold northeast, Betty J. had a long history of bronchial attacks. Often, her seizures were so severe that she would choke and gasp for air, and her face would turn blue as she looked ready to expire. Frequently recurring attacks made her moody and depressed.

Warm Climate a Dim Hope. Frequent vacations to a warm climate made her feel better, but she was still subjected to bronchial seizures and various allergic attacks. She caught colds in the warm summer climate that were just as severe as in her home in the cold winter northeast! The fault, she discovered, was with her weak cell-tissue breakdown. She had a history of eating refined, processed and dehydrated foods. She rarely touched "bunny food," which she labelled the needed plant foods. Her years of improper self-care had weakened her cell-tissue membranes, and her body was ravaged by infectious invaders. When a warm climate was a dim hope, she acceded to a friend's suggestion that she follow this simple but effective program:

1. *Breakfast:* Whole grain cereal in skim milk with fruit juice and Brewer's yeast flakes. Occasional eggs could be had throughout the week. Fruit or vegetable juices as beverages. Coffee substitute such as Postum or a herbal tea with lemon juice and honey.
2. *Luncheon:* A large raw fruit or vegetable salad seasoned with herbs and any kind of vegetable oil. Chopped fish salad. Assorted seeds and nuts. Whole grain bread slices. Raw fresh fruit for dessert. Fruit or vegetable juices as beverages. Coffee substitute or herbal tea.
3. *Dinner:* A main protein from any of these groups: lean meat, fish, cheese, peas and beans, nuts, or a bowl of cooked soybeans. A large raw vegetable salad. Dessert was fresh fruit.
4. *Do's and Don'ts:* All foods had to be as natural as possible. The preceding menu offered a balance of high-protein, moderate fat and moderate carbohydrates, and a treasure of enzymes and essential fatty acids. She was

told *not* to have any refined sweets or starches. No cakes or pastries. She was told to restrict or eliminate as many artificial foods as possible. The emphasis was upon natural foods containing the natural protein, vitamins and other elements that gave her body the working materials for better health.

REWARDS: It took three weeks of this delicious diet, consisting of good first-class protein and a balance of other nutrients, to begin the rebuilding of her billions of fragile cells and tissues. This simple but natural food program so strengthened the cellular membranes that the harmful microorganisms could not get through. Now the bloodstream used its red blood cells for transporting needed antibodies to any infection site in order to weaken, destroy and later wash out infectious wastes.

Secret Effectiveness of Protein Program: Protein and the other nutrients succeed in uprooting and casting out harmful bacteria from the intercellular spaces of the tissues, including the blood. These bacteria would otherwise create viruses that would then act as parasites, devouring cell membranes, depleting the DNA-RNA feeding action, or "starving" the cells themselves. Protein protects against the destruction of the cell's membranes, and then follows by strengthening them, and also searching out and eradicating the allergy-causing virus bacterial wastes. This is the secret or little-known infection-fighting power of protein. But more importantly, *adequate daily protein-nutrient balanced intake offers a form of natural immunity that is capable of offering good protection against virus or allergic attacks.*

Recovers, Recuperates, Rejuvenates. Betty J. responded on this good program and recovered from her lifelong bronchial seizures. Her tubules and membranes were strong. Allergens could not invade. Blood protein now banished infectious wastes. She recovered so much that she loved the cold, felt so young in the winter, and said that, thanks to protein, she was able to enjoy nature on a year-long basis—free from respiratory distress!

HOW PROTEIN CAN ENRICH YOUR BLOODSTREAM TO FIGHT INFECTION

Adequate protein intake offers your lymph nodes (small pea-sized bodies in the course of the lymphatic vessels of your bloodstream) an important infection-fighting fortification.

Protein Creates Filtering Benefit: Metabolized protein nourishes the cells and tissues of the lymph nodes. Strong cellular walls enable these nodes to act as filtering beds to remove infectious waste from the lymph before it enters the bloodstream. Protein is a "great wall" that shields out foreign and undesirable invaders from the bloodstream. Protein then nourishes the white blood cells, which are needed to ingest and destroy bacteria, decayed particulates, wastes, cellular debris and infectious wastes.

Builds Better Infection Protection: Protein enables your white blood cells to manufacture antibodies (known as lymphocytes) and to proliferate and boost production of needed antibodies. Protein nourishes your bone marrow so it can then manufacture the lymphocytes or antibodies. It then takes them to the thymus, the spleen and lymph nodes for multiplication by cell division. These substances are *all* composed of lymphatic cells and tissues, and desperately need protein for their very existence. A strong white blood cell count depends upon protein—the key to better and longer-lasting *natural* protection against infection.

The body defenses depend upon protein for cell-tissue nourishment, and for the production of those antibodies that can give good protection against infections.

PROTEIN MIST FOR BETTER BREATHING

James I. finds that a natural inhalator that contains both protein and vitamins can work miracles in opening up clogged nasal and respiratory tissues, and actually nourishes his breathing apparatus so he feels better. Equally important, James I. can *eat* the materials out of which the spray mist is prepared. It's as healthy as it is delicious.

How to Make a Protein Mist: Prepare any stove-top protein meal. A chicken or turkey soup with marrow bones and cooked vegetables is a good combination of protein with other essential nutrients. Or make a beef or liver stew. Animal foods are preferable since they need to be cooked and they offer a more *concentrated* vaporous mist.

How to Inhale Your Protein: Make a towel or foil tent over the kettle and put your head beneath. Now breath the steamy protein vapors up to twenty or thirty minutes. Keep breathing in and out deeply. Then remove the "tent" and you can now *eat* this protein food.

Benefits of Protein Mist: It has helped unclog James I.'s choked respiratory-nasal-pharynx network. It makes him breathe easier for most of the night. The cooked protein releases a store of *vaporous amino acids* that go to work instantly in helping your tissues become strengthened. These same inhaled amino acids work to rebuild your membranes and thereby reduce the effectiveness of the harmful microorganisms. James I. has such a *Protein Mist* two or three times a week and notes that his allergic reactions are subsiding. He hopes he will soon recover completely as a result of this double-action process of breathing in amino acid vapors and then eating the protein food.

The amino acids nourish the mucous membranes of the breathing apparatus and keep them healthfully moist. Well-nourished cells and tissues offer hope for natural and youthful immunity against infections.

Protein and nutrients give the body the working materials out of which new tissues and membranes can be properly rebuilt and strengthened to resist the threat of bacterial invasion. A healthful program calls for balanced food intake and good hygiene habits. It's the natural way to freedom from infections.

CHAPTER HIGHLIGHTS

1. Use vitamins with proteins to help form a *combination* to stimulate body production of antibodies, the bulwark against infections.

2. Strengthen your cellular membranes with protein fortification for resistance against bacterial invasion.

3. A set of three "vita-pro" programs offers the benefits of natural antibodies, cellular immunity and virus protection.

4. The *Vita-Pro Bronchial Healer* effectively guards against tissue breakdown; the *Vita-Pro Cold Fighter* offers cellular strength against microorganisms; the *Vita-Pro Allergy Soother* activates cellular regeneration and protects against allergic invasion.

5. Betty J. recovered from bronchial attacks on a simple food program. She is youthfully immune in her cold winter residence.

6. James I. uses a *Protein Mist* (and eats the source, too) for relief of clogged nasal and respiratory organisms. All natural!

11

HOW TO CONTROL
STRESS AND TENSION WITH
PROTEIN PROTECTION

Hypertension can be controlled with the stress-shielding action of protective protein. Recurring hypertension can produce a deficiency of protein and starve the millions of cells and tissues that might otherwise offer soothing protection against the stresses and tensions of daily living. But protein offers a *miracle* of cellular insulation and protection against the corrosive ravages of occasional or recurring tension.

Why Protein Is Needed for Stress-Tension Control. By the time the middle forties of your lifespan have been reached, your body has accumulated decades of "stress shocks" which include environment, physical and emotional disturbances, various common and uncommon illnesses and a cumulative consequence of the tensions of daily living. Years of such stress shocks have depleted your cells and tissues (of the body as well as the nerves and glands) of needed protective protein. This makes your body more sensitive to more stress situations.

Daily stresses provoke a pronounced increase in the excretion of nitrogenous substances (nitrogen being a major ingredient of amino acids, the cell-tissue building blocks of protein). The more nitrogenous substances excreted, the greater the sensitivity to tension and irritants of daily living. This outpouring of nitrogen creates a loss that may continue for several weeks or longer. Over a period of time, it may seriously deplete the stores of body protein, thereby rendering the body's cells and tissues more vulnerable to the corrosive effects of tensions. A fortification of protein acts as a natural "stop gap" against nitrogen depletion, and helps to "coat" the cells and tissues and provide natural insulation against stresses and tensions.

Maintain Adequate Daily Protein for Stress-Protection. The amount of protein you lose is directly proportional to the amount of stress your body is subjected to. This means you should have an adequate daily amount of protein for protection against stress symptoms. For example, mild stress causes a nitrogenous-protein loss for only a few days. But more serious or *prolonged* stress may cause larger daily losses of nitrogen for weeks or even months. If this protein is not made available as replacement, then the body and mind react with nervous symptoms that can be physical or emotional or a combination of both. Protein offers a natural stress-protection.

FIVE PROTEIN FOODS WITH NATURAL
STRESS-PROTECTIVE FACTORS

Metabolized protein sends a stream of amino acids to the body's cells and tissues to create a stress-protective "coating" factor, as natural insulation against the ravages of daily living. Here are five such everyday foods that are tasteful and also healthful. They help replace the "nitrogenous loss" that contributes to nervous unrest.

1. *Natural Nut Butter on Whole Grain Bread.* The non-processed nut butters (sold at health stores and many supermarkets) are made with polyunsaturated non-hydrogenated fats and offer a double bonus of protein power. Nuts are prime sources of almost all known amino acids. In the form of a butter used as a spread for whole grain bread, you have the time-release benefit created by the polyunsaturated oils in the spread. This means that when you eat a nut butter sandwich, your body is getting a steady and slowly-metabolizing amount of amino acids that help provide day-long protection against nitrogenous excretion. A lunch of a nut butter sandwich with a fresh fruit salad is a healthful way to fortify your cells and tissues with needed protein to help you cope with the stresses and strains of daily living.

2. *A Plate of Cheese Squares and Slices.* Here is a concentrated supply of almost all known amino acids. Cheese squares are good for nibbling. They provide a high biological value to create metabolic adaptations for nourishment of the cells and tissues involved in nerve responses. (The quality of a protein is frequently indicated by its biological value, which is a term used to express the percentage of absorbed nitrogen which is retained in the body for maintenance and growth.) The biological value of food proteins depends upon the amino acid proportionality pattern and quantity, as well as the

digestibility of the protein. Milk products are second (after eggs) in the highest biological values needed for good nerve and body health. You will be soothing your cells and tissues, as well as your nervous network by having a plate of cheese squares with whole grain crackers and a warm or cold beverage at least once every other day. You'll be fortifying your system with protein of high biological value. You'll feel soothed and better able to meet the stressful challenges of daily living.

3. *The Miracle Protein Breakfast for Day-Long Protection.* Whenever Peter W. has to face a day of tension-filled conferences or sales meetings, he starts off with a power-packed protein breakfast of just *two* basic foods. Here it is:

Scramble eggs in any vegetable oil. Add chunks of avocado slices just before the eggs "set." If desired, sprinkle with sesame seeds or wheat germ. Eat for breakfast.

Benefit: The egg contains a high biological value of protein containing all known protein. The avocado contains protein, too, but it offers a rare oil seldom found in fruits. This oil is in a form that contains valuable fatty acids that send a *slow and steady* nitrogenous supply to the body's cells and tissues. It is this "time release factor" that can help offer day long protection against protein depletion.

Peter W. finds that his nervous system and his brain cells become strengthened because of this breakfast. He is no longer jumpy, irritable or prone to temper loss because of stress. This simple *Miracle Protein Breakfast* gives him day long protection against cellular depletion. He has a calm nervous system. He thinks more clearly. He is able to rise up the corporate ladder of success with freedom from stress and tension, thanks to the *Miracle Protein Breakfast.*

4. *Tasty Lunch for Mid-Day Cell-Feeding Protein.* Broiled liver or any glandular meat is a prime source of needed protein. When you broil liver in wheat germ, whole grain flour or soy flour, you add complementary vitamins that help in the metabolism of the protein, and give you better amino acid nourishment. For lunch, try a plate of broiled liver and a side dish of raw fresh vegetables with a slice of whole grain bread. You will be giving yourself a treasure of cell-feeding protein to help ease that mid-day slump. It will soothe your nervous system and help you cope with the stresses of the rest of the day.

5. *A Plate of Mixed Plant Foods for Balanced Protein.* While animal proteins have high digestibility and an adequate supply of

essential amino acids, plant or vegetable proteins (which may be low in one or more essential amino acids) can be *combined* to offer a better protein quality than if taken singly. Combine any assorted seeds and nuts. Try combining natural brown rice and milk. Try a bowl of any whole grain cereal and fruit in milk. Sprinkle most of these foods with natural wheat germ. A plate of steamed beans, nuts, raw almonds and seeds, and a bowl of cottage cheese with fruit slices can give you *balanced* protein that acts as an insulator for your nerve cells and tissues, and shields you from the tensions of daily living.

Simple Protein Plan: Each day, plan to eat adequate protein in one form or another, from an animal or plant source or a combination of both. Maintain a *balance* for more complete amino acid nourishment. This helps give your body needed nitrogenous nourishment to replace that which is excreted during stress situations.

Stress can be caused by a feeling of frustration when you miss a train or stop light. Stress can be caused while waiting for an important phone call. Stress can be caused when you sit quietly and think about a simple problem. Stress can be caused by chemicals in our environment (and food). Stress is part of our civilized daily living. To meet these challenges, Nature has provided protein as an important insulator to protect the nerve cells and tissues against the stress-provoking situations of everyday living.

HOW AN EGGNOG CAN GIVE YOU PROTEIN-PLUS
FOR STRESS-FREE HEALTH

Doris K. has discovered that when she has an eggnog at noontime, she experiences a feeling of overall contentment and relaxation. More importantly, she finds that the eggnog eases her nervous twitching, the irritating "tic" of her eyelid, the frequent slurring of her words, and the annoying "jitters" that made her hands tremble when performing the simplest of tasks.

Doris K. eats a balanced meal, but because she has so many responsibilities, which include managing a household of seven (not including herself!), working part-time at a dress shop, and serving on local voluntary clubs, she is frequently "edgy" and "nervous." The problem here is that Doris K. is subjected to so many responsibilities that her protein stores are used up rapidly. They need replacing.

Special Protein-Plus Secret: Doris K. finds that an eggnog can soothe her nerves and give her clearer thinking. Her symptoms subside, even vanish. What is the secret of a simple but protein-power beverage such as an eggnog? Here it is.

The adrenal glands are involved in a hormone that responds to stress. When the body is subjected to stress, the adrenals release extra steroids which help control physical-mental unrest and provide extra energy to the body. The steroids or hormones are prime sources of protein. But even more important is that these hormones contain Vitamin C. *The adrenal glands have a higher concentration of Vitamin C than any other body organ, including the liver.* During stress, the adrenal glands secrete large amounts of *both* protein and Vitamin C.

During prolonged stress, these levels are depleted. A protein deficiency can lead to nervous unrest and physical symptoms. A Vitamin C deficiency can compound and increase these problems. Just as nitrogenous-protein substances are excreted through gland hormones during stress situations, so is Vitamin C given off and a deficiency can reduce the body's ability to withstand stress. A *combination* is needed as a natural tranquilizer to soothe the cells and tissues and replace that which is lost. This is the "secret" of the "stress insurance" benefits of a simple but powerfully effective *Protein-Plus Eggnog.*

Here is how Doris K. makes it:

Protein-Plus Eggnog

2 eggs, beaten	1 cup whole or skim milk
2 tablespoons honey	1 cup orange juice

Combine eggs with honey and beat well. Now mix in milk and orange juice. Beat until thoroughly blended. Drink slowly.

Rewards: This *Protein-Plus Eggnog* is a natural blending of protein (all known and essential amino acids) and Vitamin C. Together, they feed the adrenal glands and help replace the mixture of protein and Vitamin C lost through excretion during stressful situations.

It has helped Doris K. cope with her activities. It has given her renewed cell-tissue replenishment so that the pressures of everyday living can be met with vigor. Well-nourished adrenal glands need the *combination* of protein and Vitamin C found in the *Protein-Plus Eggnog* for use as a buffer zone against stresses and tensions.

TEN FOODS THAT ACT AS "NATURAL TRANQUILIZERS"

Here is a selection of top-notch everyday foods that are combined to offer you a blending of needed protein with Vitamin C. When eaten in a combination, or even singly, they converge to nourish your cells and tissues and help make up for that which is lost during daily activities. These two nutrients, protein plus Vitamin C, join with others to soothe the network of cells and tissues comprising the nervous system and promote a feeling of well-being. It's easy (and tasty) to eat these ten foods throughout the weeks as part of your nerve-regeneration program:

1. *Apples and cheese.* A highly beneficial combination of needed vitamins plus protein that swiftly promote soothing relaxation.

2. *Peaches and cottage cheese.* Sprinkle with wheat germ and some fruit juice for good blending of needed cell-tissue foods.

3. *Cantaloupe and diced turkey.* A joining of valuable vitamins that prompts protein metabolism and speeds amino acids to the millions of nerve endings throughout your body.

4. *Citrus fruit and chicken.* Oranges, a lemon wedge, or grapefruit wedges with cold diced chicken add up to a powerhouse of nutrients for cell-tissue rejuvenation and relaxation of nerve cells.

5. *Tomatoes and cheese.* Try stuffing hollowed tomatoes with cottage cheese (add nuts for super protein) and eat as a delicious luncheon. A treasure of needed vitamins and protein.

6. *Green pepper strips with cold meat slices.* Eaten together, your cells and tissues are rewarded with vitamins and protein for natural regeneration.

7. *Salad greens with diced eggs.* Makes a high vitamin-protein combination that is appetizing and also nourishing for the body's cells and tissues.

8. *Cauliflower with chopped liver.* Speedily after eating, you'll feel the way your body is relaxed and contented. The Vitamin C of the cauliflower merges with the complete protein in the liver and uses the amino acids for cell-tissue replenishment.

9. *Broccoli with chopped meat.* Broccoli vitamins use the protein from the chopped meat to soothe and replenish the substances lost during emotional tensions.

10. *Cabbage with broiled meat slices.* Extremely good Vitamin C that acts upon the available protein (cooked meat has more easily assimilated amino acids) to send a vital supply of substances needed for cell-tissue replenishment.

For the sake of better cell-tissue and nerve health, plan to eat a variety of these ten foods that have the nutrients to give you the feeling of a Natural Tranquilizer.

HOW PROTEIN PROTECTS YOU FROM EFFECTS OF LOW BLOOD SUGAR

Many folks "live on their nerves" and are subjected to chronic excitement, anxiety and nervous tremors. These folks keep driving themselves, doing without proper sleep and without proper food, until they notice that they reach the state of mind-body exhaustion. This is a symptom brought about by a condition of protein-deficiency known as hypoglycemia or "low blood sugar."

Protein for Youthful Vigor. Your body's source of youthful vigor is in a substance called *adenosine triphosphate,* or ATP. This substance is released by protein within your cells, and transformed into *adenosine diphosphate* and *phosphoric acid,* or ADP. The secret of energy is that *protein* is the food that transforms ATP into energy-building ADP. Protein sends this substance to the body's millions of cells, tissues and glands to produce youthful energy. Protein holds the key to youthful vigor by this transformation. If your body is adequately protein-ized, then you should not have to be a "bundle of nerves" or feel the effects of average work and activity. But if you are protein-starved, then your cells and tissues are deficient in ATP, and this creates a form of mental and physical weakness. You are tempted to drive yourself, without correcting the cause, and this may worsen the situation. It may create an emotional and/or physical breakdown. With protein, this situation could be avoided.

How Protein Protects Against Hypoglycemia. This condition is one in which there is a deficiency of needed sugar in the bloodstream. An intake of high sugars and starches causes a swift metabolism and a swift burst of energy. Once the sugar is used up, the body reacts with nervous tension, harder breathing, loss of energy. Protein spares the metabolism of carbohydrates and enables it to be slowly used for gradual energy. Protein transforms sugar into *glycogen,* a substance that is stored in the liver. Protein uses this storage substance to convert into energy during strenuous effort. Whenever your body undergoes stress (it could be as simple as worry or as serious as a body shock), protein acts as a messenger to use *glycogen* from the liver for rebuilding worn out nerve cell-tissues and rejuvenation.

Protein is needed to give the body its needed resources to protect against low blood sugar or hypoglycemia.

Protein Soothes the G.A.S. Syndrome. A bad day at the office, a difficult day with the youngsters, a family problem, even worrying over bills, causes a drain of needed body sugar. It starts what is known as the G.A.S. syndrome. Specifically, it is the *general adaptation syndrome.* Namely, first stress depletes the body of needed protein, and sugar is speedily metabolized for instant and brief energy spurts. Secondly, the body's starved cells and tissues respond with an alarm reaction. Then the body musters whatever biochemical resources it has to keep itself going. The body uses up needed nutrients. Bit by bit, body and mind functions start to decline. To help guard against the G.A.S. Syndrome, protein is needed *daily,* with a combination of moderate fat and moderate carbohydrates so that they cannot be so speedily metabolized. Instead, protein acts as a "watchdog" to see that carbohydrates are *slowly* used up to provide the body with a steady supply of youthful energy.

How Protein Gave Anna M. a "Forever Young" Feeling of Body and Mind. Anna M. was as sensitive as an exposed electric wire. Just the slightest touch and she would jump. She was the victim of frequently recurring headaches and allergic attacks. Often, she would feel her temples pounding with tension. There were nights when she would lie awake in a profuse sweating and feel "panic" as she faced the cold pre-dawn hours. She was told that there was "nothing wrong" with her. But Anna M. felt prematurely old (she was just approaching her 40's), complained she could hardly remember simple facts or faces, and that she was an "old woman in the body of a young one." It became a vicious cycle as Anna M. drove herself throughout her daily tasks by drinking sugary beverages, eating sweet cakes and pastries, and drinking gallons of sweetened coffee. This worsened her problem. Then a well-meaning friend, who had the same problem but corrected it, told Anna M. how she could protein-ize her "starved" cells and tissues and rejuvenate her mind and body. The program was simple:

FOODS FOR CELL-TISSUE NOURISHMENT

All meats and fish. Dairy products (eggs, milk, butter and cheese). All fruits and vegetables *except* those listed here to restrict.

Eat nuts as in-between meal snacks for concentrated protein.
Natural nut butters on whole grain bread.
Postum. If you must have coffee, try Sanka or caffeine-free coffee.
Most beans.

FOODS TO RESTRICT

Potatoes, corn, macaroni, spaghetti, pie, cakes, pastries, sugar, candies, dates, raisins, cola and other sweet soft drinks, processed cereals and processed grains.

Benefits: This easy program offers a high protein, moderate fat and moderate carbohydrates. Most important, *these foods enable a gradual absorption rate of carbohydrates because of their protein contact. This controls the sudden rise in blood sugar and the subsequent fall which creates cell-tissue depletion and hypoglycemic symptoms.*

Restores Nerve Health. Just seven days on this tasty program gave Anna M. a feeling of renewed youthfulness. Her mind was alert. Her tremors stopped. Her headaches and allergies subsided. She was no longer sensitive to the touch. She enjoyed healthful sleep almost every night. She was now a young woman in a young body, and said she had a "forever young" attitude toward the good days ahead. Life was enjoyable, thanks to protein-izing her nerve cells and giving her body a "watchdog" to control the G.A.S. or *general adaptation syndrome* condition. Protein nourished her body. Protein stored glycogen in the liver, and then was able to draw out *slow* amounts as needed. It was this "time release" process that made Anna M. look and feel young again.

Keep Active, Eat Protein. You can enjoy cell-tissue energy and keep active. It's good sense, as well as good health, to keep yourself vitally alert and energetic. But you should eat protein so it can "borrow" the energy-producing carbohydrates to be used to give you this energy. Protein also controls the replacement and restoration of energy-producing substances in your body's reserves. Protein controls the payment and repayment of the body's cell-tissue feeding debts. Protein controls the energy-producing ATP, which is the direct source of ADP, to protect against symptoms of stress. Protein helps you keep active by feeding your cells and tissues with needed energy-producing ingredients.

Protein Protects Against Hypoglycemia. Miracle protein controls the flow of insulin from the pancreas, regulates the uptake of glucose by your body's cells (it helps your body metabolize sugar into energy-producing substances), and controls the conversion of glucose into glycogen in the liver for storage and use as needed. This helps protect against hypoglycemia in which there is an insufficiency of necessary sugar. *By feeding the adrenal glands, protein can help maintain proper carbohydrate metabolism and relieve average symptoms of hypoglycemia.* This is the miracle protection power of miracle protein!

How to Enrich Foods with Natural Protein. You can "enrich" many foods with natural protein. For example, enrich whole grain cereals, hotcakes, muffins, cookies, waffles with soy flour, wheat germ and powdered milk. You can sprinkle unflavored gelatin powder over most cereals and fruit or vegetable salads. Sprinkle wheat germ over almost all kinds of salads. Add wheat germ to dairy foods, especially cottage cheese for super protein benefits along with most needed nutrients. Add several tablespoons of low-fat but high-protein milk powder to skim milk. You'll have a naturally protein enriched beverage with little of the fat. Add chopped eggs to most salads. Use yogurt as a filling for homemade puddings or desserts. Sprinkle chopped seeds and nuts over cereals and salads. Snack on cashew and other nuts. Sprinkle chopped nuts over a cheese and fruit salad. These are highly delicious and natural protein enrichment treats. They're every bit as tasty as they are beneficial in helping to build and rebuild your cells and tissues.

Hypertension afflicts folks who live in tranquil as well as hectic environments. It is a consequence of our modern times. Each stress situation drains away vital protein and vitamins from the cells and tissues. Prolonged depletion may cause wastage and premature aging. Stress and tension symptoms are usually the first to be noted when there is inadequate protein in the body's reserves. Feed your cells and tissues needed protein and your nerves will sing with joyous happiness. Life will be beautiful.

IN REVIEW

1. Protein offers protection against "stress shocks" of everyday living.

2. Enjoy any or all of the five protein foods that help replace the "nitrogenous loss" that contributes to nervous unrest. These five foods offer natural stress-protective factors.

3. Peter W. has a simple breakfast consisting of *two* foods. In combination, they give him a miracle protein power and he enjoys day long protection against tension-filled situations.

4. Doris K. uses an eggnog for protein-plus cell-tissue feeding. She is alert, active and free from temperamental outbursts because of this healthy and good eggnog that is made from everyday foods, probably in your pantry, in a matter of moments. Works swiftly, too.

5. Select any or all of the ten foods that are delicious as "natural tranquilizers."

6. Protein is a natural protector against hypoglycemia or low blood sugar. It helped give Anna M. a "forever young" mind in a "forever young" body.

7. You can easily enrich everyday foods with natural protein. Do it at home in a few minutes for a feeling of "lifetime youth."

12

HOW TO USE PROTEIN TO FEED YOUR GLANDS FOR "FOREVER YOUNG" HORMONES

Every cell and tissue of your body receives its regenerative nourishment through the medium of life-promoting *moisture*. Whether your cells are close to your skin surface or deep within your body, they depend upon a supply of liquids for healing repair and replication. The DNA and RNA molecules which act as "central directors" of youthful processes are especially in need of this moisture for their life-giving functions. The cellular rejuvenation process depends upon the *ability* to absorb needed moisture. This ability is made possible when the cells receive sufficient protein, which is used for "building blocks" for their walls and inner components. Feed your cells sufficient protein and they can absorb the needed amino acids that will make them strong and youthful. The key to feeding your cells protein is in having adequate hormones from your body glands.

Glands Secrete Cell-Feeding Protein Hormones. A gland is a body organ that secretes a liquid substance from its cells to the billions of body cells. A gland that does not excrete its substance but leaves it to be picked up by the bloodstream for distribution to all body parts is known as a *ductless* or endocrine gland. All glands secrete moisture that is largely protein. This moisture is used to nourish your billions of cells and tissues, and to promote regeneration from within your body. This moisture, known as a *hormone*, is largely protein. Your glands serve many functions, but the prime benefit is to protein-ize your cells and tissues and keep them youthful and healthy.

Hormones: Rivers of Youth. Simply speaking, hormones are protein liquids created by the body glands, then transported to other body organs for the purpose of extending and maintaining youthful processes of life. The proteins in the hormones create a stimulating effect. The word *hormone* is taken from the Greek language, meaning "to excite." The hormones stimulate and alert the cells by nourishing them with protein. This process makes them "rivers of youth" as they go from head to toe, nourishing, repairing and replicating your billions of body cells and tissues. Hormones offer hope for a forever young health of your body and mind through their protein feeding functions.

Hormones Are Treasures of Protein. Basically, hormones are composed of protein. For example, the adrenal gland needs protein to secrete tyrosine which is protein, too. Thyroxin from the thyroid gland is a combination of the amino acid tyrosine and iodine. Insulin secreted by the pancreas consists of *nine* different amino acids. All glands use metabolized protein as energizers so they can create hormones which are primarily protein for nourishment of the body from head to toe.

Body Glands Work in Harmony. To maintain youthful health, *all* your body glands must work in smooth harmony. *All* the glands need a regular supply of protein and other nutrients for their own health, and for the manufacture of hormones. *If one gland is protein-deficient and starts to slow down, all the others follow suit. Every one of your glands must be protein-nourished and in good working order, otherwise your entire body declines in health.*

TWO CELL-FEEDING FUNCTIONS OF PROTEIN HORMONES

When all of your glands are working in smooth harmony, they take up amino acids (metabolized protein) and use *two* different substances within these amino acids to create two specific cell-feeding functions:

1. *The "Element" Youth Factor.* Your glands send "elements," or amino acid substances traveling throughout your bloodstream, to reconstruct your tissue cells and help extend the look and feel of youthfulness.

2. *The "Ether" Youth Factor.* Your glands send "ethers," or amino acid substances, through your nerve cell network to be deposited on your brain cells for "feeding" and improvement of their capacity to give prolonged alertness and protection against senility. It is important for your hormones to contain the "ether" youth factor (a protein) because it is used to nourish your brain cells *daily*. The

reason is that the brain shrinks, as 100,000 cells die each day. The "ether" factor must be made available without deficiency in order for the brain to continue functioning.

Like a furnace thermostat which regulates the amount of fuel output according to room temperature, your hormones regulate your body organs, needs, requirements and conditions. Protein hormones may be compared to body fuel. When they are available in adequate supply, the thermostat of youthful life can continue functioning. A deficiency creates discomfort and a decline in health.

Over 40? Extend Youth with Protein Hormones. The activity of the glands slows up in the early 40's. This need not be so if the body glands are properly nourished with a *balanced* food program that calls for emphasis upon protein. The slowdown can be pepped up in the 40's and can continue to be alerted throughout many more decades of youthful life through good protein. Since hormones are made of protein, the body requires this miracle food as a means of helping to extend the prime of life. You can feed protein directly to your glands. You can give your glands the *elixir of life,* protein, through wholesome, healthy and everyday foods. Life and health can begin at 40 with protein hormones. Here is a list of the 7 basic glands, their functions, and how to use certain foods to give them *balanced* nutrition with protein emphasis and be rewarded with youthful health at all ages.

1. THE THYROID GLAND

Location: A two-part endocrine gland that resembles a butterfly and rests against the front part of the windpipe.

Protein Hormone Youth Function: The thyroid uses protein to secrete an important hormone, *thyroxin,* to stimulate the metabolism of body cells. The protein in the hormone regulates the rate at which your body consumes oxygen and burns up foodstuffs. Too little protein causes hypothyroidism, which causes decreased metabolism, a premature aging or thickness of the skin, puffy, dry skin and a slowdown of mental and physical functions. Protein-carrying thyroxin nourishes your cells and tissues and guards against these aging problems. This hormone helps extend youthful appearance, thanks to its availability of protein.

How to Feed Protein to Your Thyroid: The secret here is to use a food containing protein *plus* iodine. In *combination,* these nutrients are used by the thyroid for the manufacture of thyroxin that is transported throughout your bloodstream for nourishment and cell-tissue rejuvenation. The protein-iodine foods include egg yolk and salt-water fish, such as cod and haddock. *Suggestion:* Combine two foods such as boiled fish and garlic, because the latter reportedly has the highest iodine content of any food plant. Broiled fish is a prime source of ocean protein and iodine. Garlic is a prime source of most essential nutrients with iodine. In combination, they become dynamic feeders of protein *plus* iodine for your thyroid gland. Also use garlic in your raw vegetable salads. Try a chopped egg and garlic salad for a powerhouse of protein-iodine hormones.

2. THE PITUITARY GLAND

Location: The size of a pea, it hangs from a short stalk at the base of the brain.

Protein Hormone Youth Function: The pituitary is often called the "master gland" because its three lobes use protein to secrete at least nine known hormones. All of these hormones contain protein and need protein for stimulation of your hypothalamus (a portion of your brain). The pituitary uses protein to alert this hypothalamus to secrete a protein substance called the corticotropin-releasing factor which, in turn, promotes the secretion of the protein hormone, ACTH. (Abbreviation for adrenal cortex trophic hormone.) Protein is a major component of ACTH, which is used to regulate your metabolism and promote better resiliency of your inner muscles and organs. Protein-containing ACTH is the body's natural defense against arthritis. This protein hormone, ACTH, then stimulates the cortex of your adrenal gland to boost its manufacture of hormones such as *hydrocortisone.* This is a *natural* arthritis-fighter as opposed to the *synthetic* cortisone administered to those troubled with this problem. It appears that Nature has made protein the food and the spark plug to stimulate a natural cortisone hormone to protect against the age-causing problem of arthritis. Pituitary hormones use the protein to control the health of the body's skeleton, muscles and tendons. The pituitary hormones enter up into the rhythm of cellular processes and convert the end products into healthful, youthful vitality.

How to Feed Protein to Your Pituitary: A balanced *combination* of animal and plant protein appears to nourish the pituitary so it can promote the secretion of its minimum of nine hormones. Many of these secreted hormones contain *combinations* of animal and plant protein. Feed this pituitary a *balance* and it will be able to stimulate the flow of cell-tissue feeding hormones. A plate of cooked soybeans, steamed in vegetable oil, together with a serving of cottage cheese offers a good balance of animal-plant protein. Chopped liver and onions, together with a side dish of assorted seeds and nuts, is a deliciously healthy way to nourish your pituitary with needed hormone-making protein. Baked chicken or turkey, together with a dish of cooked lentils or peas, offers another tasty combination of pituitary-feeding protein. Sprinkle sesame seeds over scrambled eggs and enjoy with whole grain bread and yogurt, with wheat germ added for a good "brain boosting" protein hormone breakfast. It helps alert your brain, and repairs cells and tissues speedily so you can have better thinking power for the day ahead.

3. THE ADRENAL GLANDS

Location: Shaped like Brazil nuts, you have a pair of adrenal glands, sitting astride each kidney.

Protein Hormone Youth Function: Known as the "glands of emergency," they use protein for the manufacture of *adrenalin* or *epinephrine.* The adrenals take amino acids and use them to nourish the cells and tissues of its two portions, the outer cortex and the inner medulla. Then these two portions promote the secretion of adrenalin, which is needed to spur every nerve and muscle into split-second and perfect coordination. For example, when you turn the brakes on in an emergency, when you leap out of the path of an oncoming bus, or when you must race into or out of a building during an emergency, your adrenals are signalled through the billions of cells lining your nervous system. Then the adrenals speedily pour out the needed hormone to help you cope with the emergency. The adrenals need protein for their own hormone manufacture. Your body needs protein to nourish the nerve cells, or there will be poor "emergency hormone" and poor "nervous transportation" along your nerve system.

Protein hormones issued from your adrenals help your nerve cells become more alert to stimuli; they work to help you see better, to hear better, to regulate and strengthen your heartbeat, to improve your powers of respiration, to strengthen your muscular system. Protein hormones such as adrenalin perform this function. They also help to neutralize toxic wastes in your bloodstream to help you recover from ailments more easily, and to help build a more youthful cellular network from head to toe. As a bulwark against stress, adrenal protein hormones are absolutely essential.

How to Feed Protein to Your Adrenals: Protein and Vitamin C work together to create the stress-shield hormones. They also combine to guard against hypoglycemia or low blood sugar. It is known that during stress or hypoglycemia, there is a great loss of Vitamin C. It appears that the replacement of this Vitamin (which is taken up by protein and then stored in the cortex) offers a combined protection against such problems as fatigue, weakness, poor appetite, irregular weight and hypertension. Combine protein foods such as lean meat, fish, eggs, cheese, peas, beans and nuts with citrus fruits that have Vitamin C, such as oranges, grapefruits, strawberries and lemons. A meal consisting of protein foods and a dessert of Vitamin C foods adds up to a welcome combination that feeds protein to your adrenals and feeds youthful vitality to your cells, tissues and brain. It is the natural protein hormone way to better body health.

4. THE PARATHYROIDS

Location: There are four of these glands; each is about the size of a pea. They are located at the four corners of the thyroid, against the front part of the windpipe.

Protein Hormone Youth Function: When protein is metabolized, its amino acids go to the parathyroids, which use them to secrete a hormone known as *parathormone.* It helps regulate the body's use of calcium and phosphorus. The protein is used by the *parathormone* to facilitate the transportation of calcium from the skeletal structure, then into the bloodstream and the cells and tissues. This brings about a regeneration of these tissues. The parathyroids, themselves, are in need of a *combination* of protein and calcium. *Both* of these nutrients tend to regenerate the cells and tissues of the glands. Once this is accomplished, they use remaining protein plus calcium to create the needed parathormone, and send calcium throughout the body for nourishment of the billions of cells and tissues, via the bloodstream. A deficiency of protein and calcium may cause insufficent hormone flow, and this may contribute to nervous instability, mental distress and outbursts of uncontrollable rage. The glands use protein and calcium to cleanse the toxic substances within the bloodstream. They neutralize some of the corrosive effects and thereby enable the parathyroid hormone to perform its cell-tissue rebuilding without interference.

How to Feed Protein to Your Parathyroids. The secret is to combine protein with calcium for better effectiveness in cell rebuilding. This was discovered by Rudy Q., a retired machinist, who had become the terror of his family due to a parathyroid problem. He was given to frequent outbursts of temper upon the slightest provocation. Often his hands would shake, as if he had palsy. (Calcium deficiency is responsible for the trembles.) Rudy Q. was difficult to reason with. He was stubborn. He became loud-mouthed. While his food program was nourishing, there was a deficiency of calcium. This created an *imbalance* so that his parathyroids were deficient in needed minerals. Protein helped regenerate some of the cells but they needed calcium for better rebuilding. When Rudy Q. made a simple improvement, he began to recover. His disposition sweetened. His hands were steady. He developed clearer thinking abilities. He was well-mannered and appeared youthful. The simple change called for an increase in his use of dairy foods for needed minerals, especially calcium. Rudy Q. would have a lunch consisting of skim milk, cottage cheese and fruit slices, sprinkled with wheat germ. Another day, he would sprinkle chopped nuts into a cup of fruit-flavored yogurt. Or he would enjoy a vegetable-cheese casserole as a meal in itself. This gave him needed calcium. Daily, he would eat one or two protein foods such as lean meat, salt-water fish, eggs (the whites are pure protein and can be eaten freely by cholesterol-watchers since they contain no cholesterol), assorted peas, beans and nuts. With an increase of minerals such as calcium, and with protein, Rudy Q. enjoyed better mental health. The billions of brain and body cells could now be nourished by the *parathormone* which, itself, is composed largely of calcium and protein. Rudy Q. had gained a new and youthful lease on life, thanks to this simple food plan.

5. THE PANCREAS

Location: A large, long gland (called the sweetbread in animals) located behind the lower part of the stomach. It has a "head" and a "tail."

Protein Hormone Youth Function: When this gland receives protein, it helps transform it into amino acids that then secrete the needed hormone known as insulin. Amino acids in insulin work to help the bloodstream convert body sugar to its needs. Amino acids influence insulin to *burn* the sugar. A deficiency of protein means that the cells and tissues of the pancreas break down. The gland now becomes weak and it cannot secrete a balanced amount of insulin. Sugars and starches cannot become properly *burned* and the threat of diabetes follows. It was this situation that plagued Claire McD. who had a craving for sweets, starches and excessive fats, day in and day out.

The problem is that overeating of such foods *displaced* the desire for eating protein. The insulin-producing cells of the pancreas became "starved" and could not be repaired sufficiently because of a deficiency of protein. Claire McD. experienced overweight, excessive thirst, slow wound healing, brittle nails that easily split and many skin blemishes. Since she had a family history of diabetes, Claire McD. sought out help from specialists who told her to make some amazingly simple adjustments to help "wake up" her sluggish pancreas.

Claire McD. gave up refined sweets and starches. In their place, she ate fresh fruits and vegetables which gave her *natural* carbohydrates. But this meant her taste buds could now appreciate protein foods. When she ate protein foods, together with fruit and vegetable carbohydrates, her digestive system could use the metabolized amino acids to nourish the cells and tissues of her pancreas. Then this gland could use the protein to maintain a balance of the insulin hormone so it could work more effectively in burning sugars and starches. Protein acted as a natural *regulator* so that the insulin was *balanced* and could more efficiently screen out and filter sugars and keep a healthier bloodstream. Soon, Claire McD. no longer felt abnormally thirsty, her skin cleared up, she developed skin with the bloom of a young girl, and her fingernails became strong and clear again. She slimmed down, too. Claire McD. avoids refined sugars and starches, eats wholesome fresh fruits and vegetables, lean meats, poultry, fish, scrambled egg whites, assorted cheeses from skim milk, seeds and nuts. *Protein plus fresh plant carbohydrates act as healthful pancreas cell feeders.* They can repair the cells and tissues of the pancreas (and body, too), so that insulin hormones (which contain both protein and carbohydrates) can efficiently metabolize sugars and starches and keep your body in youthful working order.

How to Feed Protein to Your Pancreas: A combination of protein with plant carbohydrates acts favorably in cell-tissue replenishment. Your protein meal should include a serving of any fresh or cooked fruit or vegetable. Taken together, you'll be able to nourish your pancreas with needed protein and natural carbohydrates which help balance a healthful insulin supply.

6. THE THYMUS

Location: Situated on the windpipe just below the thyroids. While large at birth, it slowly shrinks until adulthood, when its influence declines. Yet it still remains an important gland in need of protein nourishment.

Protein Hormone Youth Function: This gland is nourished by protein which uses this miracle food for the metabolism of such minerals as calcium and phosphorus. The thymus hormones transport protein to the brain for cellular replenishment. This helps promote better thinking. It protects against premature senility. The thymus hormones use protein for the manufacture of white blood corpuscles that guard you against infection. The thymus hormones use protein for building and replenishment of red blood cells, too. This helps promote better healing. This gland is influenced by the pituitary. This indicates the importance of overall body nourishment by protein for balanced health and youthful vitality.

How to Feed Protein to Your Thymus: Good grain protein from wheat germ, cereals and whole grain products appears to offer good cell-tissue rebuilding processes to the thymus. The thymus hormones contain a form of protein which is likened unto grain protein. Feature whole grain breakfasts several times weekly. Sprinkle with wheat germ and add fresh fruit slices. Your digestive system will use these vitamins and protein for overall body nourishment yet also for thymus gland cell replenishment. You'll feel more mentally alert and more resistant to infection too.

7. THE SEX GLANDS

A. Ovaries or Female Sex Glands. Location: In the lower region of the female. They make possible the process of fertility and reproduction.

Protein Hormone Youth Function: Protein is meeded by the female sex glands for the manufacture of *estrogen* and *progesterone,* the two female secretions that give her youthful fertility and a youthful appearance. These hormones, themselves, are largely protein and are made from supplies of protein in the cells and tissues of the ovaries. If there is a deficiency, then the ovaries become depleted and their cells break down. There should be a ready availability of protein for nourishment of the sex glands so that the female can enjoy a youthful appearance in body and mind, too. These female hormones use protein to give her a smooth skin, a better feeling of youthful vitality and, in short, make her into the lovely woman Nature intended her to be.

How to Feed Protein to Your Ovaries: A good Sex Gland Tonic can be made easily. In a glass of fresh fruit juice, stir two tablespoons of desiccated liver powder (sold at all health stores) and add a tablespoon of honey for better taste. Stir vigorously. Drink just one glass a day. This Sex Gland Tonic is a rich powerhouse of protein plus vitamins and minerals that helps nourish the body's cells, and more importantly, the female ovaries, and it replenishes destroyed cells. The metabolized amino acids of the Sex Gland Tonic then become part of the youth-building hormones, *estrogen* and *progesterone.* Also eat good protein foods such as lean meats, poultry, cheese, skim milk dairy foods, beans and nuts.

B. *Prostate or Male Sex Gland.* Location: Just below the bladder, encircling the urethra where it exits from the bladder. The prostate surrounds the urethra (the urinary pipe that extends from the bladder to the elimination organ) like a "mitten" around a pipe.

Protein Hormone Youth Function: The prostate needs protein to manufacture *testosterone* within the testes in the scrotum. This youth hormone helps protect against premature aging. It is the youth hormone that sends protein throughout the body so that muscles become firm, so that the brain functions with alertness, so that the male can function with virile vitality. With the approach of the 40's, if there is a deficiency in protein and other nutrients, the prostate slows its production of the youthful testosterone; as a consequence, its cells and tissues wither. But instead of shrinking, the prostate enlarges. This swelling may also accompany infection. It calls for having to get up nightly, to make increasingly more frequent bathroom trips throughout the day. It may include pain, as well. While prostate removal ends this problem, it also ends the production of *testosterone*, the hormone that promotes male youthfulness. It also ends his powers of fertility. Some men become sexually useless, too, when subjected to this operation, and this results in a combination of mental and physical conflict.

How to Feed Protein to Your Prostate: The male seminal fluid and the prostatic fluid hormone both appear to be largely composed of protein. To nourish this essential "male youth gland" so that it can continue to send forth youth-building hormones, your daily diet should include lean lamb or beef, lots of fresh and cooked vegetables and many fresh fruits. In particular, you should have a lot of raw seeds and nuts, as these are used by protein to keep the prostate gland "cleansed" and to nourish the cells and tissues involved with the manufacture of the hormones. Avoid heavy starches, sweets and volatile spices, as these act as *irritants* to the prostate and can inhibit as well as destroy formation and rebuilding of needed cells and tissues. Cleanse and soothe the mucous cellular membranes of the prostate with lean meats, poultry and enzyme-rich raw fruits. This combination is favorable for replenishment of the prostate gland's cells and tissues, and can help extend the youthful prime of life in males. The male hormones are basically constructed of protein so they need a daily supply of this essential nutrient.

PROTEIN: NATURE'S MIRACLE HORMONE FOOD

A plant withers and dies when it lacks water, which is considered a hormone food for a plant. Likewise, the human body cells can wither and die if deficient or deprived of needed water or hormone foods. Protein is Nature's miracle food. It protects against premature aging by giving hormone "moisture" to the cells. This enables the cells and tissues to absorb the protein. At the same time, the hormones help cells wash away toxic wastes or clogged debris. This improves the sponge-like qualities of the body cells. This helps replenish them so they can keep replicating and help extend the lifespan.

The body cells and tissues look like those of a sponge. Keep them protein-moisturized through hormones, and they can reproduce and function adequately. The body's glands can also continue their hormone-making functions that promote youthful health.

But if the body cells are deficient in protein-moisture, they become "thirsty." They atrophy. They absorb less and less of needed protein. The cellular walls break down. They dry. They shrivel. The reaction is noted on a body that is inadequate and weak. Age symptoms include a parched and wrinkled skin and flabby flesh. Just as protein-starved cells dry up and lose life, so is your life shortened. The answer is to feed protein to your body so that your glands can become cellularly strong and issue forth protein-ized hormones, the miracle life-promoting moisture.

SUMMARY

1. Glands are made of protein. Their cell-feeding hormones are largely protein.

2. Protein hormones have two basic cell-feeding functions. They contain "elements" and "ethers" that influence youthful cellular regeneration.

3. Seven glands influence the foundation of extended youth. Each gland needs protein. Each gland issues protein-containing hormones.

4. Rudy Q. was rescued from senility through an easy protein hormone program calling for several everyday foods eaten throughout the week.

5. Claire McD. was able to ease her sugar-starch obsession and protect herself against an erratic insulin flow by improving her protein intake. Her thirst eased up, her skin was blooming, she slimmed down. Her pancreas could now provide her protection against diabetes.

6. Protein acts as a natural moisturizer to the glands and the body; it nourishes the billions of cells and tissues. Protein is Nature's miracle hormone food.

13

THE "BALANCED PROTEIN" WAY TO A SLIM-TRIM-HEALTHY SHAPE

Since your body structure is composed largely of protein, a reducing or weight-maintenance program should be aimed at shedding pounds or inches but *not* protein. Your billions of cells and tissues depend upon protein to keep your muscles, organs, skin and just about every part of your body in youthful condition. A reducing program that is *low* on protein causes a depletion of this needed substance from the cells and tissues, and body breakdown may result. If there is a prolonged protein deficiency in a slimming program, then the body has to burn its own muscle tissues for its protein needs. This creates the wrinkled skin, the unsightly lumps and bumps, and the scrawny look that often accompanies low-protein slimming programs. *It is unwise and unhealthy to skimp on protein or any food element when slimming down.* The key to healthy slimming down is to maintain a "balanced protein" program in harmony with other essential foods.

YOUR 10-STEP "BALANCED PROTEIN" SLIM-SHAPE PROGRAM

This program gives you a balance of all essential elements and emphasizes protein so that you can feast your way to a slimmer shape. It gives you the joy of eating with good nutrition while you shed inches or pounds. It's the fun way to lose weight.

1. Use the following as desired.

bouillon	herbs	seltzer
clear soup (fat-free)	horseradish	spices
club soda	lemon and lime	tea
coffee substitute	paprika	vinegar
		water

2. Eat all you want of the following, cooked or raw, vegetables.

artichokes	carrots	spinach

asparagus	cabbage	cucumber
beans, green, snap or waxy	cauliflower	dandelion greens
bean sprouts	celery	endive
beet greens	chicory	escarole
brussels sprouts	collards	kale
lettuce	chinese cabbage	kohlrabi
mushrooms	pickles	leeks
mustard greens	pimentos	squash, summer
okra	pumpkin	squash, winter
onions	radishes	swiss chard
parsley	rutabaga	tomato
parsnips	sauerkraut	turnips
peppers	scallions	turnip greens
		watercress

3. The following vegetables and cereals should be used more sparingly than those listed under 2.

corn (fresh or canned)	grits	lentils (fresh or canned)
dried beans or peas	macaroni	Sweet potatoes or yams
green limas	noodles	brown rice
green peas	plantain	whole wheat spaghetti
cornmeal	potatoes	

4. *You may eat three servings daily of any unsweetened fruit that's fresh, canned or frozen. Choose at least one Vitamin C fruit daily. This vitamin combines with protein to help build cells and tissues without adding unnecessary weight.*

apples	grapefruit juice	peach
apricots, fresh	honeydew melon	pineapple
blackberries	lemon	plums
blueberries	lime	raspberries
cantaloupe	orange	strawberries
cranberries	orange juice	tangerines
grapefruit	papaya	

Use more sparingly:

apricots, sun-dried	guava	pomegranate
bananas	mango	prunes, sun-dried
cherries	nectarine	raisins, sun-dried
dates	pear	watermelon
grapes	persimmon	

(*Note:* Most canned and frozen foods are sweetened; be sure to drain off heavy syrup before using the food.)

5. Eat whole-grain breads or cereals.

6. Limit eggs to 2 or 3 a week. Cook in the shell, poach or scramble without fat.

7. Include two cups of skim milk daily. Fat-free buttermilk or evaporated skim milk may be substituted for good protein intake.

8. Broil, pan broil, simmer, bake or roast protein foods such as meat, fish or poultry. Do *not* fry! Remove all visible fat before eating. Do *not* eat gravies or sauces. Eat at least three seafood meals weekly for good protein intake. Limit fat group to three times weekly.

LEAN GROUP

bass	halibut	shad roe
bluefish	herring*	sturgeon, fresh
butterfish	kidney	tongue
carp	liver	trout
chicken	mackerel	tuna fish*
cod	pike	turkey
finnan haddie	salmon*	veal
flounder	sardines*	weakfish
haddock	shad	whitefish

*Canned fish with added oil should be drained well.

FAT GROUP

beef, lean	frankfurter	liverwurst
bologna	lamb, lean	

9. Avoid alcohol; avoid excessive amounts of rich gravies, cream dressings, fat spreads, candy, pastries and carbonated beverages.

10. Exercise adds to the success of your diet program in several important ways. It stimulates the assimilation of protein to help tone muscle tissues and to increase the protein-carrying blood supply to nourish your many vital organs and tissues. Exercise also helps burn up calories that might otherwise be stored as fat in the body. Good exercises include walking, bicycle riding, swimming, bowling or any doctor-approved athletic activity. Plan to activate your body with physical movement at least one hour per day. It helps stimulate better protein metabolish while burning up excess calories.

This 10-step "balanced protein" slim-shape program lets you enjoy good foods and gives you a balance of all needed nutrients, with emphasis upon protein. You can slim down while enjoying tasty foods on this easy-to-follow program.

THE "MIRACLE PROTEIN REDUCING TONIC"

Marcia O'D. loved to eat. She was always hungry. After she would finish an extra-heavy meal, she would start to nibble and nibble, until she would finish the snack box. Then Marcia O'D. would go back to the refrigerator and eat leftovers. She was a bulgy *un*pleasingly plump woman who looked much older than her later 30's. A double chin, unsightly fat from her stomach and buttocks, and chunky thighs all made her the local scapegoat. People snickered when she walked down the street. Youngsters rudely called her "Fatty" or "Chubby" or else they'd yell out, "Here comes the big tub of lard!" It made Marcia O'D. break down in tears of frustration. How could she control her runaway appetite?

Secret Tonic with Three Ingredients. At a local waist watching meeting, she heard a "formerly fat" woman tell that she heard of a secret tonic calling for three ingredients that could put a stopgap on a runaway appetite. This woman lost over 120 pounds and was now a slim 126, because she *controlled* the obsessive craving for food. The easy-to-prepare tonic was described by this ex-fatty. It excited Marcia O'D. and she prepared it.

"Miracle Protein Reducing Tonic." In a glass of fruit or vegetable juice, dissolve one tablespoon of unflavored gelatin. Dissolve a meat bouillon cube (sold at health stores or most food markets) and then mix together. Drink this "Miracle Protein Reducing Tonic" about one hour *before* regular mealtime. It has a unique ability of putting a brake on your eating urge. You will find yourself having a reduced eating urge. You'll find yourself reducing, too.

Sheds Unsightly Pounds. Marcia O'D. followed this simple program at once. Results? She found that her appetite was controlled. She ate half portions of everything and was fully satisfied. The nibbling urge was eased, then totally eliminated. Marcia O'D. shed over 70 pounds while following this easy program. Soon, she was down to a fetching 116. Gone was the unsightly double chin. Gone were the heavy bulges hanging from her arms. Her torso was as lean as that of a mannequin. Now when she walked on the street, she was whistled at, instead of riciculed. She calls the "Miracle Protein Reducing Tonic" the million dollar slimming secret. It makes her look and feel like a million dollars!

Secret of "Miracle Protein Reducing Tonic." The fruit or vegetable juice contains vitamins, minerals and enzymes which are used by the *complete* protein in the combination of gelatin and beef bouillon to give your cells and tissues a healthy feeding and saturation. Digestive cells and tissues are now "plumped" up and will not give you the hunger pangs ordinarily felt on a low protein-fat program. The secret is to "feed" protein to your digestive cells without gaining weight. This is possible in the low-calorie but high-protein and high-nutrient "Miracle Protein Reducing Tonic." Just drink one glass before each meal. The *complete* amino acid pattern actually "drenches" and "moisturizes" your digestive cells and this gives you a feeling of comfortable fullness. It's the natural and tasty way to control your appetite and help yourself slim down, thanks to protein satisfaction.

HOW TO USE PROTEIN TO BUILD CELLS WHILE LOSING EXCESS WEIGHT

Many overweight folks discover that when they lose weight, they develop unsightly lumps, bumps or sagging flesh. Many have stretch marks. This haggard or wan appearance is often traced to a low protein intake on a reducing program. Cell-tissue degeneration has caused a breakdown beneath the skin so that there is premature wrinkling or sagging of the surface skin. This is known as "cellulitis" in which a gel-like substance made up of fat, water and wastes is trapped in immovable pockets beneath the skin. This calls for a high-protein, moderate or low-fat and low-sugar program.

Why Protein Can Control Cellulitis. Metabolized protein, in the form of amino acids, offers a slow and steady metabolism of fat cells so that nitrogenous action can help "flush" out excess fat from the body. A *deficiency* of protein means that melted fat cannot be metabolized or "evaporated" and becomes trapped in cell pockets. This causes the unsightly bulges, wrinkles lumps and "scarf skin" or "gizzard" throat that occurs with improper diet programs. Protein can control *cellulitis* by helping to metabolize the liquefied fat and then sloughing it out of the body through normal eliminative channels.

A natural law of maintaining a slim figure or losing excess weight is to fortify your body's cells and tissues with adequate protein in a balanced diet. Protein will help protect against *cellulitis* or scrawny, bulgy fat lumps and creases that make reducing less than the joy it should be.

Here's how to use protein to nourish your cells and promote excess fat loss:

1. *Protect against hunger.* Protein has a more lasting effect upon hunger pangs. Try a tablespoon of non-hydrogenated nut butter. You'll have a little fat with good plant protein.

2. *Healthy nibbling.* Stuff a celery stalk with any nut butter and munch away. This helps ease appetite pangs. It gives you needed protein for cell feeding, too.

3. *Appetite Control.* In a glass of any liquid (skim milk is good because it has low-fat but good protein) stir one tablespoon of Brewer's yeast. You'll be able to skip coffee breaks and nourish your cells with this appetite-controller. It's loaded with almost all known vitamins, minerals and prime protein. Very tasty, too.

4. *Protein for dessert.* Avoid calorie-building sweetened desserts but satisfy your dessert desire with a good protein dessert. Have a fresh fruit salad with cottage cheese. Or try fruit-flavored gelatin for good protein with vitamins. Try a small portion of brown rice pudding flavored with cinnamon. A plate of assorted seeds and nuts is good. You have to chew very well and you satisfy your jaw and throat muscles and feel less inclined to overeat. High protein helps stabilize your blood sugar and you'll feel fewer hunger pangs, real or imagined.

5. *Poultry is good cell-building food.* Turkey and chicken offer a high-protein of top quality and good biological value so that your cells and tissues become firmed up while the deposited fat can be sloughed off during metabolism. Plan to eat turkey several times a week to satisfy your "meat urge" and still enjoy a low-fat and low calorie food. Turkey and chicken are among the *leaders* in high cell-building protein, while being among the *lowest* in calories and saturated fat. Nibble or munch on a poultry drumstick as a healthy snack. Feed your appetite and your cells while you slim down.

You need a regular, *daily* amount of sufficient protein so that your billions of cells and tissues can be repaired and renewed, while you are losing weight. It's essential to get protein daily, since it is not stored in the body for long. A deficiency will compel your body to feed upon itself. This breaks down tissue and muscle. This causes fat to be trapped since protein oxidation cannot remove it if unavailable. This creates a haggard *cellulitis* appearance of wrinkling and aging. This could be protected against with delicious protein, every day, in a balanced food program.

HOW TO PROTEIN-IZE YOUR THYROID GLAND
FOR A SLIM-TRIM SHAPE

Grace N. had tried one diet after another. Nothing worked because she could not stop eating. She would devour four extra-heavy meals a day. In between meals, she would gobble up pizzas. Grace N. was a corpulent 238. She kept on gaining because she kept on eating. She might have continued eating her way to a shortened life, until she read how protein could actually feed the thyroid gland and put a stopgap on the appetite. Grace N. tried a very simple remedy. It worked. She found her appetite declining. Soon, she eliminated eating between meals. Then she ate smaller and smaller mealtime portions. Gradually, she enjoyed three small meals daily. Her weight melted down to a slim 134. She had gone from a size 24 to a size 16. She was still losing weight, thanks to a simple protein-ization of her thyroid gland.

Secret of Weight Control with Thyroid Gland Control. Thyroxin, the protein hormone issued by the thyroid gland, determines how swift the body will burn up food calories. A well-nourished protein-ized thyroid will issue a protein-ized hormone that permits healthful metabolism of foods. But a starved thyroid issues faulty hormones and body fat can accumulate. Thyroxin is weak because of insufficient protein. It cannot properly metabolize foodstuffs. The thyroid gland requires protein plus iodine, a mineral with which it combines to issue the thyroxin hormone that helps maintain healthful weight. In the situation of Grace N., a protein remedy fed her thyroid so that the hormone could then perform its normal fat-melting functions and help protect against obesity.

Protein + Iodine = Reducing Hormone. The all-natural remedy used by Grace N., which she called her "Protein Slim-Shaper," can be prepared this easy way:

Protein Slim-Shaper

In a glass of vegetable juice, stir one-half teaspoon of kelp (a seafood powder sold at almost all health stores and special food shops) and one tablespoon of desiccated liver. Stir vigorously. Drink one glass in the morning. Drink another glass in early evening.

Secret of "Protein Slim-Shaper": The protein of the liver is energized by the enzymes and minerals of the vegetable juice. The protein takes the iodine from the kelp, metabolizes it, and then transports it through the bloodstream to your thyroid. The gland now makes thyroxin, a hormone that is stored in sacs or follicles within the gland. They are protein-bound when released into the bloodstream, to physiologically step-up cellular metabolism throughout your body. This influences the brain's hypothalamus, which monitors the need for thyroxin. When your appetite-center (in your hypothalamus) signals an eating urge, then your hypothalamus signals your thyroid to issue the needed hormone. This rhythm appears to put a control signal on your appetite. It's an all-natural appetite-controller that feeds your cells and tissues and promotes better tissue respiration. It gives a wonderful feeling of youthful well-being, while helping to keep you in a slim shape.

The iodine of kelp is prime food for your thyroid, and it requires protein for better assimilation; therefore, the "Protein Slim-Shaper" contains the desiccated liver. It's a Nature-balanced reducing tonic. It's all-natural!

Bonus Tip: To keep your thyroid gland well-nourished so that the thyroxin hormone can send valuable protein to your appetite control center, you should plan on serving ocean fish for an iodine-rich protein food. You'll be able to *fill up without filling out!*

PROTEIN PLANNING GUIDE FOR SLIMMINGLY YOUTHFUL HEALTH

Here are ways in which you can plan on using protein to keep your cells and tissues and your body in slimmingly youthful health.

Breakfast: Soften several tablespoons of whole grain oatmeal overnight in a little water. Next morning, add fresh fruit slices and some chopped nutmeats for a protein-packed breakfast.

Luncheon: Chicken or turkey soup, beef broth or a vegetable salad with cottage cheese or yogurt and nuts, adds up to a good protein-packed luncheon. Add better nutrition to soups with a cupful of chopped celery. A slice of rye bread with a wedge of natural cheese boosts the cell-feeding protein power of this meal.

Protein Salad: On a raw green salad, add assorted cheese wedges, seeds and chopped nuts. Sprinkle with wheat germ.

Protein Dessert: Arrange a plate of raw fruit slices. Add nutmeats and sprinkle with Brewer's yeast. *Chew* thoroughly to satisfy this part of the eating urge and you'll find yourself less inclined to keep nibbling after mealtime. TIP: Thorough chewing builds up the supply of enzymes so that protein can then be better metabolized and assimilated by your cells and tissues.

Protein Drink: Use a blender to combine skim milk, sunflower meal, several cut up fruit slices and a banana. Whirr together and then drink. You can use this as a meal-in-itself. It's high in protein, but very low in calories and fat. It's very good for nipping your appetite and sending streams of protein moisture to your thirsty cells and tissues.

Protein Pep Drink: If you feel weak because you've cut down on heavy meals, try this beverage to give you speedy vigor. Combine two tablespoons of blackstrap molasses with a cup of fruit juice. Add a teaspoon of vegetable oil. Now toss in a handful of dates or other fruit. Add an egg yolk. Blend or whirr together until fully assimilated. This is a delicious beverage that is brimming with protein. It provides speedy energy and is a healthful pickup.

"Protein Break": In a cup of chicken broth, blend an egg yolk and several teaspoons of chopped chives. Whirr together. Drink as a substitute for a coffee break. It has "staying power" and will help make you energetic and less inclined to eat for many, many hours.

Super-Protein Booster: If you need something extra-special in helping to soothe your stomach pangs, and want to get speedy protein that is assimilated rapidly, try this all-natural remedy for a tired feeling spawned by reducing: In a cup of skim milk (or pulverize raw nuts or seeds for a protein powder which is mixed with water for high-protein seed milk), add avocado chunks, a dollop of yogurt, wheat germ and a banana. Whirr together. Drink slowly. You'll feel the protein going to work almost at once as a sense of contentment takes over. Your compulsive appetite will be less commanding. You'll feel remade as, indeed, this is the task of the protein of this special booster.

THE CELL-FEEDING PROTEIN WAY TO A SLIM-TRIM FIGURE
Daily Eating Program

FOOD	PROTEIN	CALORIES
Non-fat milk, 1 pint	17.0 grams	170
Whole milk, 1 pint	17.0 grams	330
High-protein foods, 3 servings		
Low-fat cottage cheese, ½ cup	19.5 grams	100
Egg, 1 (limit to 2 to 4 per week)	6.5 grams	70
Beans or low-fat meat alternate	7.0 grams	95
Whole-grain cereal, 3 servings		

Chart 3

Oatmeal, 2/3 cup	3.5 grams	100
Whole-grain bread*, 2 slices	4.0 grams	115
Dark-green leafy or yellow vegetables, 2 servings		
Broccoli, 1 cup	5.0 grams	45
Leafy green salad, ¾ cup	2.0 grams	20
Whole orange or grapefruit half	1.0 grams	75
Seasonal fresh fruit: Strawberries, 1 cup, or cantaloupe, ½, or apple, 1 small	1.0 grams	55
Oil or equivalent in margarine, dressings, etc., 4 teaspoons	.0 grams	155
	66.5 grams	1,000

*1 medium potato may be substituted for one slice.

Chart 3 (Continued)

A balanced food program with proper attention paid to protein can do much to keep you looking and feeling slim, trim and healthy. It is much easier to keep excess pounds off than to take them off. So use the benefits of protein toward that goal: keeping pounds off for good health from top to bottom!

IN REVIEW

1. Protein can help keep you slim. Follow the delicious 10-step "Balanced Protein" Slim-Shape Program. It'll make you look and feel slimmingly good all over.

2. Marcia O'D. used a "Miracle Protein Reducing Tonic" and lost over 120 bulgy pounds within a reasonable time. She looks and feels like a million dollars!

3. Protein can control *cellulitis,* the problem of wrinkled, scrawny skin, and the consequence of crash or unwise dieting.

4. Protect yourself against wrinkled skin from weight loss by following the 5-step list of protein planners.

5. Grace N. controlled her compulsive eating by protein-izing her thyroid gland. She used an easy "Protein Slim-Shaper" that had only three everyday ingredients but provided her with the appetite-control she needed to slim down.

6. A variety of protein programs can easily be followed so you can enjoy slimmingly youthful health.

14

FRUITS, VEGETABLES, HERBS AND OTHER PLANT PROTEINS FOR MIRACLE REJUVENATION

Your body cells and tissues become better regenerated when nourished with protein foods coming from *the bottom of the food chain*. The DNA-RNA molecules are able to replicate strongly and vigorously with a balance of plant foods. When your cells and tissues are protein-nourished with foods from the *bottom of the food chain,* they appear to live longer and offer greater protection against decay and aging.

How Cells Are Nourished by Protein from Bottom of Food Chain. To begin with, some solar energy is captured by green plants through the process of photosynthesis. Animals (at the top of the food chain) begin by consuming these plants and liberating some of this energy for their own use. Other animals, in turn, consume some of these animals. The progression of the flow of energy through this system is known as a food chain. It starts with plant-eating animals, then goes lower to animal-eating animals, then descends lower to man-eating animals.

At each transfer of energy in a food chain, a high percentage of the biological energy stored in the organism of the lower level becomes unavailable to those of the higher level.

By moving down the food chain, much more energy becomes directly available. In other words, the further down the food chain man acquires his food, the stronger is the Nature and solar-created protein for optimum cellular replication and DNA-RNA nourishment.

By eating a variety of plant foods, you should be giving your cells the more powerful form of protein as created by Nature.

Plant Protein Considered First Class for Cell Rebuilding. Large amounts of first class edible protein are fed to animals who often inefficiently convert it into protein that will be eaten by humans who eat meat. It is believed that when you eat the plant foods normally eaten by the animals, *you will be feeding your cells first class protein as created by Nature.* When you eat animal foods, you feed your cells second class protein which has first been used by the animals. For good biological cellular nourishment, your program should contain an abundance of plant foods.

Less Pesticide, Fewer Chemicals. By eating lower on the food chain, using more plant proteins, you also reduce the quantity of pesticides and chemical residues in your system. Animal foods are chemically treated and preserved. But plant foods are more natural and, if organic, are free of additives. Even if not organic, you can wash and peel many plant foods and lower the amount of pesticide residue. In contrast, meat foods are injected with chemicals which cannot be removed. Since your cellular molecules are destroyed by chemicals and additives, it is essential for rejuvenation to control the amount of pesticide infiltration in your body. Plant foods offer such a possibility.

The following chart shows the protein content of plant foods.

MEATLESS PROTEIN FOODS[1] –
OFFICIAL U.S. GOVERNMENT LISTINGS

(All portions are 100 grams, approximately 4 ounces or one-half cup.)

FOOD	GRAMS PROTEIN
Soybean Flour (low Fat)	43.4
Brewer's Yeast	38.8
Peanut Butter	27.8
Wheat Germ	26.6
Peanuts	26.0
Cheese (Cheddar)	25.0
Sunflower Seeds	24.0
Almonds	18.6
Sesame Seeds	18.4
Cheese (Cottage, uncreamed)	17.0
Eggs	12.9
Dried Split Peas (cooked)	8.0
Cow's Milk (Fluid)	3.5

Chart 4

[1]Watt, B.K. & Merrill, A.L. *Composition of Foods:* Agricultural Handbook No. 8, United States Department of Agriculture, Washington, D.C., 1963, pp. 6-67.

FOOD	GRAMS PROTEIN
Soybean Milk (Fluid)	3.4
Brown Rice (Cooked)	2.5
Avocados	2.1
Olives (Green)	1.4

Chart 4 (Continued)

HOW YOUR BODY REJUVENATES ITSELF WITH PLANT PROTEINS

The process of digestion occurs when your enzymes break down large molecules of plant foods into smaller molecules that can be absorbed by your small intestine. Therefore, your stomach and intestine break up proteins, the largest molecules known, into their constituent amino acids. Your enzymes then enable these plant amino acids to pass through the walls of your intestine into the rest of your body. Your billions of body cells pick up these plant proteins, then rebuild them into new cells and tissues and nourish the DNA-RNA elements. This is the "rhythm of rejuvenation" made possible by an abundance of these plant proteins.

Plant Proteins Offer More Efficient Rebuilding. Plant foods contain fewer additives than meat foods. (In most meats, chemicals and additives are lodged in the fat and muscle and cannot be removed. In contrast, plant foods can be scrubbed, scraped, peeled and washed so there will be much less chemical abuse.) This means that your body cells can use the amino acids of the plant foods with better assimilation. Plant proteins can create more efficient rebuilding.

The digested food additives cannot be absorbed into the cells. Additives and drugs, once absorbed, have no place to go. Therefore, the chemical additives of meat foods begin to accumulate in the various organs and may cause cellular disintegration and interference with cellular rebuilding. Your body will benefit from a larger amount of plant foods with a lesser amount of animal foods. This maintains a good balance and helps rebuild cells with a minimum of internal cellular drug abuse caused by chemical corrosion of needed tissues.

Miracle Protein: Key to Cell Rejuvenation. The *secret* of the superiority of plant protein for cell rejuvenation lies in this biological process. The digested protein in the form of an amino acid locks on to a particular "receptor" site on the surface of your cell. This sets off a form of biological chain reaction that relays the "message" into the interior of the cell. Replication or cellular rejuvenation now progresses at a satisfactory rate. Plant protein appears to have this miracle of cellular rebuilding while animal protein would perform this function in a slower progressive rate. This again suggests that Nature meant for your cells to receive protein from its *source* or its *foundation*—namely, at the *bottom* of the food chain, from the soil. This appears preferable to obtaining protein, second-hand, from animals. A balance is indicated.

EIGHT CELL-REJUVENATING PLANT FOODS AND HOW TO USE THEM

How can you feed your cells their needed plant proteins? Here are some suggestions:

1. *Powdered Milk.* Health stores sell powdered soy milk which you can prepare by mixing a tablespoon or two with water. Many stores sell ready-made soy milk, too. Add it to baked goods, soup, stews and beverages, or use it as a beverage for good protein nourishment.

2. *Soybean Protein.* This is as close to meat protein in completeness as it is possible to obtain from plant sources. Use soybean products regularly. Add soybean flour in baking bread, for pancakes or waffles. Try soy grits (these are partially cooked, cracked soy beans) that can be sprinkled over breakfast cereal for nut-like taste and cell-feeding protein value, too. They can also be added to baked goods to boost protein supply.

3. *Wheat Germ.* A natural source of good plant protein. Use it as a breakfast cereal. Sprinkle it on salads, use it in soups, casseroles or when making main dish foods. Mix it in yogurt, too.

4. *Nuts.* Whether you eat them out of the shell or in the form of nut butters, nuts and peanuts are an excellent source of top-notch plant protein. They are a good source of essential unsaturated fatty acids, too, as well as many vitamins and minerals that work with protein for body cellular rebuilding. Nut butter sandwiches make good high-protein foods.

5. *Seeds.* These are prime sources of good protein and other needed nutrients. Try sunflower seeds, sesame seeds or pumpkin seeds for snacks, or mix with your salads, in your baking or with your breakfast cereals. A rich source of needed plant protein.

6. *Brewer's Yeast.* A great source of high quality protein. *Its amino acid pattern makes Brewer's yeast almost comparable to that of meat in protein structure.* Sprinkle over salads, in stews and soups;

mix it in baked goods. Sprinkle over breakfast cereals or mix it in most beverages. It's the natural way to feed your cells meatless protein.

7. *Grains.* All grains, like other seeds, are good sources of high-quality protein. In order of their potency: defatted corn germ, wheat germ, brown rice, wheat gluten, whole nuts, barley, whole corn, whole rye and buckwheat flour.

SUPER-PROTEIN: The darker the buckwheat flour, the more protein it contains. This grain is easy to grow, and needs little fertilizer. Because it is resistant to disease, buckwheat is rarely sprayed with herbicides or insecticides because it grows so swiftly that it smothers any weeds growing around it. *Buckwheat contains about 11 percent protein.* So include buckwheat as a regular food for your cell-tissue rebuilding. It's available as groats, as buckwheat flour and in cereals. Health stores and almost all food markets carry buckwheat and buckwheat products. Tasty and healthful.

8. *Legumes.* This category includes beans, peas and lentils. Economical, they contain considerable protein as well as complementary minerals such as iron, phosphorus and the B-complex vitamins. These work in harmony with protein to nourish your cells and body. Use beans as a vegetable dish. Shape cooked beans into burgers. Add wheat germ, soy four, whole grain breadcrumbs for good protein nourishment. Also use cooked beans as part of a salad with fresh raw vegetables. Serve with a homemade dressing.

In the quest for better cell-tissue replication, plant proteins should become a necessity instead of just an occasional change of pace. It's part of the plan for biological values which calls for a balance of animal and plant proteins, with emphasis upon plant proteins. While it is true that plant foods are still likely to contain residues unless they are organically grown, you can no longer say that meat automatically enjoys the advantage. *There is no way to peel a chicken or a steak as you can an apple, banana or carrot in order to get rid of the chemical that clings to the skin.* This suggests an important balance with the emphasis on plant foods.

THE SECRET "PROTEIN X" FACTOR IN PLANT FOODS

Generally speaking, animal proteins contain all of the amino acids in proportions that can be efficiently used by the body, *if* they are completely utilized. For many folks, animal foods may be indigestible with the consequence that the complete protein is not fully utilized. Therefore, the body may not receive all of the required amino acids.

Plant Protein X. This is a type of protein which is "secret" in the sense that its working method is not yet fully understood by researchers and food technicians. It is called "Protein X" because it appears to sustain the plant from its very inception until its maturity. A plant does *not* eat animal foods for its sustenance. Yet, it thrives and flourishes in good health. It possesses its own protein which is available for human consumption. This unique type of protein is self-sufficient. It is known as "Plant Protein X" and possesses complete amino acids to sustain the plant. It may well have that same miracle life-giving power for humans.

How to Tap the Cell-Feeding Power of "Plant Protein X." At each meal, try to eat portions that come from different parts of the plants. For example, try to combine leaves, seeds, stems and roots of different plants at one meal. *Benefit:* Each part of the plant appears to have a different amino acid balance. *By eating all the parts at one time, you tap the cell-feeding power of "Plant Protein X" that is close to being of the highest quality and is healthfully balanced.*

Suggested Sample: To build *whole* health, feed yourself *whole* plant proteins. Eat the whole plant, rather than just one part. For example, make a combination of green leaves (lettuce), roots (parsnips), seeds (corn or wheat) and fruit (apples, berries). Singly, such foods may have incomplete amino acid balance. But a combination eating program, containing leaves, roots, tubers, seeds and fruits offers you a biological completeness with adequate amino acids from "Plant Protein X."

How Philip R. Eats Complete Plant Proteins: A problem of severe, recurring stomach pains and excessive digestive acid made it difficult for Philip R. to eat many animal foods. He suffered severe stomach spasms whenever he would eat steak and potatoes. He would have knife-like shudders going through his abdomen when he would eat lean meat or eggs. Plainly speaking, his digestive tract had been so abused that its cells and tissues had been depleted, offering him little protection against hydrochloric acid which came pouring out whenever he ate a heavy animal meal.

He made a very simple change. He eats meat only once a week. The rest of the week, he has this simple program:

Eat nuts, seeds, avocado slices, and whole grains *together* with raw, green leafy vegetables. This gives him a complete plant amino acid pattern that is comfortably and soothingly accommodated by his digestive tract. His digestive cells and tissues are slowly being regenerated and rebuilt by the plant protein factors. His nervous stomach spasms are subsiding. His shudders are easing up. He feels comfortable. He is slowly recovering and hopes he will soon be able to eat more animal foods.

THE GRAIN PROTEIN THAT RESTORED THERESA'S VITALITY

Theresa DeM. was aging too rapidly. Her eyes were blurred. Her skin was wrinkled prematurely. She was developing sagging skin. She awakened tired. She was a bundle of nerves. A half hour of simple housework and she would go into a slump and have to lie down. No matter how many naps she had, she was always fatigued and easily exhausted. She had dark circles under her glazed eyes. Theresa DeM. looked double her age! When her sister Clara came for a visit, and saw how she was aging, she took matters into her own hands. Clara was older, yet looked younger than Theresa because she followed a balanced protein program with other nutrients. But Clara used a simple, everyday grain food for breakfast. It gave her a "super-charge" of power and vitality. She made the food for Theresa DeM. The food was *millet*.

Almost Perfect Grain Protein. Millet is a grain food that is almost perfect in its amino acid balance. It lacks *tyrosine*, but if you add wheat germ to your millet cereal, you will be feeding yourself a complete protein with all essential amino acids.

Unusual Alkaline Food. Millet is high in minerals and its amino acid pattern is alkaline so it counteracts acid and is soothingly comfortable. (While wheat, rye, oats and most grains are acid--forming, millet stands alone in being an alkaline grain.)

Protein Soothing Factor. Millet (as with sesame and sunflower seeds) contains a form of protein that will "bind" up with its minerals and alkalis and then be able to neutralize and overbalance its acid-forming constitutents. Millet protein may well be considered Nature's own healer for gastrointestinal inflammation and ulceration.

Restores Theresa's Vitality. Each and every morning, Theresa DeM. was given a bowl of millet. It would be sprinkled with abundant wheat germ or seeds and diced nuts. A few fruit slices added a naturally sweet flavor. It also provided Vitamin C for better protein assimilation. Throughout the day, Theresa ate lots of raw fruits, vegetables, and assorted seeds and nuts. Salads were made with seasonal plants and cottage cheese. Avocado slices (the avocado is a good source of plant protein) were featured in most of the salads. She had some lean poultry or chicken, if desired. Beverages were herb teas or a coffee substitute such as Postum. Within six days, Theresa became supercharged with vitality.

Her skin firmed up. She slept soundly. She awakened with alertness and eagerness for the day ahead. Her nerves were calm. She could sail through a day's work with no premature fatigue. Life was now worth living. Soon, she looked younger than her older sister.

Plant Protein X in millet and other plant foods gave her new hope for the beautiful life ahead of her.

Miracle Cell-Rebuilding Power of Millet Protein: This grain has a substance known as *nitriloside* which is taken up by protein and then used to build and rebuild cells and tissues. Millet protein is also one of the most balanced and least allergenic of all other grains. It is the *only* known grain which is able to support animal life (humans are part of the animal classification) and development when used for a longer period of time. While it is unwise to subsist *entirely* on millet, it is wise to include this protein seed grain in your food program at least several times a week.

How to Use Millet: Boil it as a cereal. Use it for breads. Add to stews, or as a stuffing for a main dish; you can mix it into patties. *Protein Meal Pattern:* Cook millet with other vegetables (in milk, if desired, for extra protein balancing) for a soup or a stew. This is a meal in a plate. Have a tossed salad with fresh fruit for dessert. You offer your billions of cells a well-balanced, high protein plant food.

BUCKWHEAT—THE MASTER PLANT PROTEIN

You plan to eat less meat and more plant foods. You may not want soybeans. You know you need protein. Where can you get it in a grain? The answer is in *buckwheat.*

Protein, Vitamins, Minerals. Buckwheat is a prime source of high protein together with thiamine, riboflavin and such minerals as iron, magnesium and manganese. *It reportedly has all the food values of* *good beefsteak, yet costs so much less and is Nature's alternate to meat protein.* It is immune to disease, requires little fertilizer, and is rarely sprayed because it can "smother" weeds in swift growing. Buckwheat has a higher tolerance for acidic soils than any other grain. It absorbs soil minerals which tend to enhance and increase its potency of protein, vitamins and minerals.

Master Plant Protein Source. Buckwheat has between 11 and 15 per cent protein. This gives it more protein than whole grain wheat and even cooked soybeans. As for *quality*, buckwheat protein reportedly comes closer to animal protein than any other grain crop.

Buckwheat protein has between 5 and 7 per cent of the amino acid lysine, which is more than high-lysine corn. *Proteins in buckwheat flour, without having any amino acid additions, reportedly have about the same nutritive efficiency in proteins of animal origin with high biological value.* NOTE: The darker the flour, the more protein it contains. Always select dark buckwheat flour and products made from this protein grain. It gives you top-notch master protein.

How to Use Buckwheat: Boil in water or milk and eat as a breakfast cereal, sprinkled with wheat germ and fresh fruit slices. Or use it in soups, for puddings and breadbaking. Mix cooked buckwheat with chopped eggs, if desired, for a solid breakfast. Use it as a stuffing. Use it for stews, casseroles and as a substitute for breadcrumbs when they are needed for recipes.

Buckwheat Honey: The very dark, reddish-brown buckwheat honey is a prime source of protein and other nutrients. The protein plant which gives you buckwheat grains also is an ideal source of honey made by bees on the farms. Use buckwheat honey for a sweetener and enjoy the good taste of protein.

Buckwheat is available in just about every food market; it is also known as groats or kasha. Buckwheat honey is sold in many food markets and health food stores. Both of them are sources of "master protein" in plant foods.

HERBS: NATURE'S GRASS PROTEINS

A cornucopia of amino acid patterns abound in herbs and spices, known as Nature's grass proteins. Herbs are more than just culinary delights to the eye, palate and digestive system. Herbs are rich in soil-cultured protein along with vitamins, minerals, enzymes, and many unidentified but necessary substances for good health.

Herbs reportedly are known to help clear up sore throats, boost cell-tissue health of the digestive system, nourish the cells of the body's glands, and even enter into the protein formation of youth-boosting hormones. Herbal proteins help cleanse the bloodstream, help the body resist infection, and promote better elimination by healing the cellular membranes of the intestinal canal and connected body organs. Herbs are unique because they reportedly have medicinal qualities combined with developed protein patterns of a superior biological value. This gives them good taste as well as good protein healing.

The art of herbology has flourished for nearly 5,000 years. In past days, when modern chemical medicines are unknown, the ancients would use herbs for healing with high rates of success. The protein in the herbs, undoubtedly, exerted a cell-tissue rejuvenation that promoted natural healing. Throughout the hundreds of centuries, herbs and spices have survived as Nature's healers from the fields and woods.

Today, we recognize that herbs are Nature's grass proteins, and when used regularly, can do much to help your body's billions of cells and tissues become renourished, replenished, regenerated and refreshed. Today, herbs are sold at most health stores and many large food markets. Chart 5 shows you how to use herbs for a variety of flavorful treats:

THE NUTRITIONAL COMPOSITION OF HERBS AND SPICES

Herbs and spices can play an extremely valuable role in good protein nutrition by helping to increase the appeal and appreciation of foods that are nutritionally important to you.

As shown in Chart 6, herbs and spices have good protein values as well as a highly beneficial assortment of other nutrients that work in *harmony* to promote good cell-tissue and general body health.

Low Calories. These analyses should be particularly good news for calorie watchers because they confirm that even the highest calorie spice (poppy seeds) will not add more than 2 or 3 calories per serving in normal usage. Considering the amount used, the others typically contribute no more than one calorie a serving and usually less.

Having this definitive data at last, all those interested in good protein rejuvenation may utilize herbs and spices to their full potential in making wholesome foods as excitingly delicious as they are good for you.

Spice Chart

HOW MUCH SPICE TO USE: When trying a new idea, it is safest to start with ¼ teaspoon of spice (excepting the red pepper spices) to a pint of sauce, soup or vegetable or a pound of meat, fish or fowl.

ABOUT PEPPER: Our most important spice deserves special attention. So versatile is its flavor, it could play a welcome role in any dish on this chart (except desserts). Good cooks always remember pepper as a seasoning corrector, often adding a final dash to taste regardless of the other seasonings used.

SPICE	APPETIZER	SOUP	MEAT and EGGS	FISH & POULTRY	SAUCES	VEGETABLES	SALAD & DRESSING	DESSERTS
ALLSPICE	Cocktail Meatballs	Pot Au Feu	Hamsteak	Oyster Stew	Barbecue	Eggplant Creole	Cottage Cheese Dressing	Apple Tapioca Pudding
BASIL	Cheese Stuffed Celery	Manhattan Clam Chowder	Ragout of Beef	Shrimp Creole	Spaghetti	Stewed Tomatoes	Russian Dressing	
BAY LEAF	Pickled Beets	Vegetable Soup	Lamb Stew	Simmered Chicken	Bordelaise	Boiled New Potatoes	Tomato Juice Dressing	
CARAWAY Seed	Mild Cheese Spreads		Sauerbraten		Beef a la Mode Sauce	Cabbage Wedges		
CINNAMON	Cranberry Juice	Fruit Soup	Pork Chops	Sweet and Sour Fish	Butter Sauce for Squash	Sweet Potato Croquettes	Stewed Fruit Salad	Chocolate Pudding
CAYENNE	Deviled Eggs	Oyster Stew	Barbecued Beef	Poached Salmon Hollandaise	Bearnaise	Cooked Greens	Tuna Fish Salad	
CELERY Salt and Seed	Ham Spread (Salt)	Cream of Celery (Seed)	Meat Loaf (Seed)	Chicken Croquettes (Salt)	Celery Sauce (Seed)	Cauliflower (Salt)	Cole Slaw (Seed)	
CHERVIL	Fish Dips	Cream Soup	Omelet	Chicken Saute	Vegetable Sauce	Peas Francaise	Caesar Salad	
CHILI Powder	Seafood Cocktail Sauce	Pepper Pot	Chili con Carne	Arroz con Pollo	Meat Gravy	Corn Mexicali	Chili French Dressing	
CLOVES	Fruit Punch	Mulligatawney	Boiled Tongue	Baked Fish	Sauce Madeira	Candied Sweet Potatoes		Stewed Pears
CURRY Powder	Curried Shrimp	Cream of Mushroom	Curry of Lamb	Chicken Hash	Orientale or Indienne	Creamed Vegetables	Curried Mayonnaise	
DILL Seed	Cottage Cheese	Split Pea	Grilled Lamb Steak	Drawn Butter for Shellfish	Dill Sauce for Fish or Chicken	Peas and Carrots	Sour Cream Dressing	
GARLIC Salt or Powder	Clam Dip	Vegetable Soup	Roast Lamb	Bouillabaisse	Garlic Butter	Eggs and Tomato Casserole	Tomato and Cucumber Salad	
GINGER	Broiled Grapefruit	Bean Soup	Dust lightly over Steak	Roast Chicken	Cocktail	Buttered Beets	Cream Dressing for Ginger Pears	Stewed Dried Fruits
MACE	Quiche Lorraine	Petite Marmite	Veal Fricassee	Fish Stew	Creole	Succotash	Fruit Salad	Cottage Pudding
MARJORAM	Fruit Punch Cup	Onion Soup	Roast Lamb	Salmon Loaf	Brown	Eggplant	Mixed Green Salad	
MINT	Fruit Cup	Sprinkle over Split Pea	Veal Roast	Cold Fish	Lamb	Green Peas	Cottage Cheese Salad	Ambrosia
MUSTARD Powdered Dry	Ham Spread	Lobster Bisque	Virginia Ham	Deviled Crab	Cream Sauce for Fish	Baked Beans	Egg Salad	Gingerbread Cookies
NUTMEG	Chopped Oysters	Cream DuBarry	Salisbury Steak	Southern Fried Chicken	Mushroom	Glazed Carrots	Sweet Salad Dressing	Sprinkle over Vanilla Ice Cream
ONION Powder, Salt, Flakes and Instant Minced Onion	Avocado Spread (Powder)	Consommes (Flakes)	Meat Loaf (Instant Minced Onion)	Fried Shrimp (Salt)	Tomato (Powder)	Broiled Tomatoes (Salt)	Vinaigrette Dressing (Instant Minced Onion)	
OREGANO	Sharp Cheese Spread	Beef Soup	Swiss Steak	Court Bouillon	Spaghetti	Boiled Onions	Sea Food	
PAPRIKA	Creamed Seafood	Creamed Soup	Hungarian Goulash	Oven Fried Chicken	Paprika Cream	Baked Potato	Cole Slaw	
PARSLEY Flakes	Cheese Balls	Cream of Asparagus	Irish Lamb Stew	Broiled Mackerel	Chasseur	French Fried Potatoes	Tossed Green Salad	
ROSEMARY	Deviled Eggs	Mock Turtle	Lamb Loaf	Chicken a la King	Cheese	Sauteed Mushrooms	Meat Salad	
SAGE	Cheese Spreads	Consomme	Cold Roast Beef	Poultry Stuffing	Duck	Brussels Sprouts	Herbed French Dressing	
SAVORY	Liver Paste	Lentil Soup	Scrambled Eggs	Chicken Loaf	Fish	Beets	Red Kidney Bean Salad	
TARRAGON	Mushrooms a la Greque	Snap Bean Soup	Marinated Lamb or Beef	Lobster	Green	Buttered Broccoli	Chicken Salad	
THYME	Artichokes	Clam Chowder	Use sparingly in Fricassees	Poultry Stuffing	Bordelaise	Lightly on Sauteed Mushrooms	Tomato Aspic	

Chart 5

APPROXIMATE COMPOSITION OF GROUND SPICES, PER TEASPOON

Spice	Wt./tsp.	Water	Food Energy	Protein	Fat	Total Carbohydrate	Fiber Carbohydrate	Ash	Calcium	Phosphorus	Sodium	Potassium	Iron	Thiamine	Riboflavin	Niacin	Ascorbic Acid	Vitamin A value
	Grams	Mg.	Cal.	Mg.	Mg.	Mg.	Mg.	Mg.	Mg.	Mg.	Mg.	Mg.	Mg.	Mcg.	Mcg.	Mcg.	Mcg.	Int'l Units
Allspice	1.7	150	6	102	112	1265	367	71	13.6	1.87	1.36	18.7	127	1.70*	1.02*	49	666	25
Basil Leaves	1.1	70	3	131	40	679	225	184	23.1	5.17	0.44	40.7	471	1.65	3.52	76	674	3
Bay Leaves	1.3	60	5	99	114	980	328	48	13.0	1.43	0.26	7.8	693	1.30*	5.46	26	606	7
Caraway Seed	1.8	110	8	385	416	783	187	101	12.6	9.0	0.40	34.0	153	6.84	6.84	146	216	10
Cardamon Seed	1.8	150	6	176	52	1336	178	85	5.4	3.78	0.18	21.6	209	3.24	4.14	41	216	3
Celery Seed	2.4	120	11	434	547	1051	310	245	43.2	13.20	4.08	33.6	1078	9.84	11.76	106	413	16
Cinnamon	1.7	170	6	78	37	1357	345	59	27.2	0.85	0.17	6.8	70	2.38	3.57	32	677	3
Cloves	1.7	90	7	107	247	1170	189	85	1.9	1.87	4.25	20.4	161	1.87	0.68*	5	1375	3
Coriander Seed	1.4	90	6	172	274	791	441	74	11.2	6.16	0.28	16.8	83	3.64	3.22	45	168	2
Cumin Seed	1.6	100	7	283	381	714	146	123	14.4	7.20	2.56	33.6	765	11.68	6.08	40	275	3
Dill Seed	2.1	140	9	275	376	1184	435	126	33.6	4.41	0.21	23.1	248	8.82	5.88	59	252	4
Fennel Seed	2.1	130	8	199	210	1277	388	281	27.3	10.08	1.89	35.7	233	8.61	7.56	126	252	22
Garlic Powder	1.5	80	5	262	9	1099	28	48	1.5	6.30	0.15	16.5	52	10.20	1.20*	10	180	3
Ginger	1.6	110	6	138	102	1158	94	91	1.6	2.40	0.48	22.4	181	0.80*	2.08	30	192	3
Mace	1.8	80	10	146	698	830	86	41	3.6	1.98	1.26	9.0	203	6.66	10.08	22	216	3
Marjoram	1.2	80	4	150	82	773	200	116	30.0	2.76	1.32	16.8	872	3.48	3.84	49	618	7
Mustard Powder	1.5	50	9	475	639	277	28	60	4.5	11.85	0.15	10.5	124	9.75	6.75	127	331	3
Nutmeg	1.5	80	11	141	739	899	59	38	3.8	3.80	0.19	7.6	42	6.84	4.75	179	228	3
Onion Powder	2.1	100	8	223	17	1690	134	73	6.3	6.09	0.84	21.0	46	8.82	1.26	12	309	4
Oregano	1.6	130	6	187	102	1038	176	144	27.2	3.20	0.32	27.2	853	5.44	0.64*	99	192	16
Paprika, imported	1.9	120	7	300	192	1102	369	184	3.8	5.13	0.38	45.6	439	7.60	22.23	293	923	64
Paprika, domestic	1.9	150	7	294	198	1146	365	144	3.8	5.70	0.38	45.6	439	11.40	25.84	291	1117	93
Parsley Flakes	1.1	40	4	241	62	597	96	155	13.2	3.41	5.94	39.6	159	1.87	13.53	87	4312	12
Pepper, Black	2.3	190	9	239	235	1529	230	106	9.2	3.68	0.23	27.6	391	1.61*	4.83	18	276	4
Pepper, Chili	2.1	140	9	294	296	1222	328	151	3.8	6.72	0.21	44.1	208	12.39	34.86	298	1338	129
Pepper, Red	2.1	130	9	336	325	1140	546	168	2.1	6.72	0.21	44.1	208	10.92	19.53	286	617	74
Pepper, White	2.2	210	9	273	176	1518	99	22	4.4	3.30	0.22	2.2	152	0.44*	2.86	4	264	4
Poppy Seed	2.5	90	13	577	887	755	217	187	40.0	0.84	0.25	20.0	247	17.25	4.50	22	300	4
Rosemary Leaves	1.2	70	5	54	209	777	228	72	18.0	0.81	0.48	12.0	396	6.12	0.48*	12	736	2
Sage	.9	50	4	92	127	561	144	69	16.2	1.96	0.09	9.0	246	6.75	3.06	51	358	21
Savory	1.4	130	5	99	73	979	214	122	30.8	1.9	0.28	15.4	529	5.18	0.56*	57	168	12
Sesame Seed	1.9	60	9	450	426	895	716	68	1.9	15.01	0.57	7.6	118	15.2	2.47	108	228	6
Tarragon	1.4	60	5	339	102	721	95	172	18.2	4.34	0.98	44.8	500	3.50	18.76	125	168	2
Thyme	1.5	110	5	102	69	1024	364	198	31.0	3.0	1.18	14.1	2025	7.65	6.0	74	180	3
Turmerie	1.9	110	7	163	169	1328	131	129	3.8	4.94	0.19	47.5	902	1.71*	3.61	91	946	3

* Present in trace amounts. Note: Mg. = Milligrams (0.001 grams); Mcg. = Micrograms (0.000001 grams).

Chart 6

Protein is the miracle food of which you are made! Plant protein offers a special type of amino acid balance with a unique biological pattern that helps promote good tissue regeneration. Plant proteins have good digestion powers. For folks who are cautious about uric acid, the end products of animal proteins, an increase and replacement with plant proteins would do much to help promote a feeling of well-being and youthful vitality. Get your protein *first-hand* from the plants, themselves, and enjoy the rewards of cell-tissue miracle rejuvenation.

IMPORTANT POINTS

1. Plant protein is considered first class for cell-tissue rebuilding.

2. You have a variety of at least eight cell-rejuvenating plant foods for daily use and enjoyment.

3. Plant foods have a secret "Protein X" Factor for better health.

4. Philip R. soothed stomach pains and excessive digestive acid by following a simple plant protein program.

5. Theresa DeM. felt a restoration of vitality, thanks to the almost perfect grain protein, millet. Her skin firmed up. Life became beautiful.

6. Buckwheat is the master plant protein. It reportedly is comparable to good beefsteak in its amino acid balance.

7. Herbs are Nature's grass proteins. Consult the chart for an endless variety of tasty and healthy ways to use herbs and spices in everyday eating. A new adventure in good tasty health with each idea.

15

HOW EUROPEANS USE PROTEINS FOR LASTING YOUTH

Visitors to Europe always express admiration at the hardy youthfulness of the peoples of many of those countries. Rosy cheeks, physical agility well into the mature age of 70 or more, the ability to keep mentally active in advanced years, and a general feeling of youthfulness are the images of Europeans received by most visitors from other parts of the world. In contrast, Americans may possess more wealth and affluence, but show a decline in youthful appearance, and cannot compete with the hardy physical endurance of Europeans, who are older in years but younger in body and mind. What are some of the youth secrets of the Europeans? Let us tap their hidden goldmine, prime the pump of knowledge, and discover how we can easily use their health programs for improved health and a feeling of lasting youth.

Maintain Protein Balance. One basic foundation supports the structure upon which youthful health is built. Namely, this is to *maintain a high protein, moderate fat and moderate carbohydrate balance.* Europeans have always known that miracle protein, in *balance* with other nutrients, can work wonders in plumping up body cells and tissues, in repairing damage, and sparking the processes that cause replication of the DNA-RNA substances.

This protein balance further helps maintain better weight control, as is evidenced by slim Europeans who may eat heartily, but are able to remain free from excess overweight as Americans know it. The secret here is that protein offers only 4 calories per gram. Fat offers 9 calories per gram. But carbohydrates offer a lower 4 calories per gram. The key to better health is to emphasize protein in foods, as do the Europeans, so that the body's fat cells and tissues are not overburdened or increased, as would be the consequence of too much fat and excessive carbohydrates. Europeans follow this simple balance and enjoy more lasting youthful health. Here are some of their reported programs.

HOW THE SWEDISH USE FOODS TO ENJOY LASTING YOUTH

The Swedish life style could be called a "health culture." In addition to participating in a broad range of exercise programs sponsored by the government for every age group, including children, young adults, office and factory workers, housewives and executives, the Swedish people follow a sensible, wholesome protein approach to diet and foods.

Here are a few hints from the Swedish Institute of Health[1] to help improve your feeling of lasting youth by using balanced protein in your food program.

Eat for the Energy You Need: In Sweden, people are schooled in health science at an early age. They know that calories are the measure of energy in every food. Fats supply the largest number of calories, almost twice as many as carbohydrates and proteins, for example. An overabundant intake of high calorie goods is unwise for a physically inactive person—which most of us are today—because unused energy calories are stored by the body as unattractive, unhealthy fat.

Balance: The Swedes are educated to watch their calories. They program themselves to either *lower caloric intake* or *increase daily exercise* to maintain proper weight. They also balance their daily diet with emphasis upon higher protein for good nutrition.

Example: All foods contain a certain number of calories as well as essential nutrients. Swedish people get the energy they need by eating foods *high in proteins, vitamins and minerals,* but *lower in carbohydrates and fats.*

Natural Foods. They believe in eating seasonally fresh, natural foods rich in protein and other nutrients, but avoiding those that are artificially and chemically processed.

Protein Energy Foods: The Swedish Institute of Public Health recommends eating balanced meals that contain foods from at least four of the following seven categories: (1) green vegetables; (2) root

[1]*Parker Natural Health Bulletin,* Vol. 3, No. 11, May 28, 1973. Parker Publishing Company, Inc., West Nyack, New York 10994. Available by subscription.

vegetables; (3) fresh fruits; (4) milk and cheese; (5) whole grain breads and cereals; (6) meat, fish, poultry and eggs; (7) butter, margarine and vegetable oils.

Important: These food categories offer a balanced amount of the protein, vitamins, minerals, enzymes, carbohydrates, fats and calorie energy they supply, making it important to *balance* and *vary* them often to maintain a naturally nutritious diet that offers sufficient cell-tissue feeding protein. This helps fuel the body metabolism and promotes a longer-lasting form of energy and youthful vitality. *And remember to take portions equal to your daily activity and the energy you expend.*

PROTEIN BREAKFAST IN SWEDEN

Since it is well-balanced, a high-protein breakfast helps energize and repair cell-tissues normally broken down during a night's sleep. The slower burning protein offers steady fuel or cell-tissue repairs that help promote a longer-lasting feeling of vitality. A typical Swedish protein breakfast might include a platter of sliced meats or cold fish, boiled eggs, yogurt, cheese, fresh fruits, juices and milk, or the daily bread of Sweden known as "crisp bread."

High-Protein Breakfast: Some Swedes crumble crisp bread (also known as rye crisp and sold in almost all U.S. food markets and health stores) into a bowl, and add fresh berries or sliced fruits with milk. This adds up to a delicious high-protein breakfast, offering a balance of vitamins, minerals, enzymes, and some carbohydrates and fat. It is nourishing food for the body's cells and tissues.

PROTEIN LUNCHEON IN SWEDEN

A Swedish lunch menu is varied, and lighter than breakfast. For work or school, lunch will include sliced or whole vegetables (cucumbers, carrots, tomatoes), cheese, slices or fine spreads of meat and fish, vegetable soup or fish chowder.

High-Protein Sandwich. A favorite high-protein luncheon is an open-faced sandwich on rye crisp bread. An apple, orange or pear adds the sweet touch to a Swedish protein lunch. The enzymes in the fruit can then metabolize the protein to usable amino acids that can be better used by the cells and tissues for youthful restoration and duplication.

PROTEIN DINNER IN SWEDEN

The Swedish dinner is the time for the well-known smorgasbord. A literal translation of this word is "bread and butter table," but there is more to it. A smorgasbord may consist of a variety of foods, hot and cold, from which one makes his own choice. A delectably light but high protein dinner might offer fish, tiny meatballs or lean sliced meat. Slices of crisp breads are used as a base for sandwich spreads. Asparagus, mushrooms and raw salads are popular. A dish of home-made applesauce is a good vitamin-mineral-enzyme dessert that will be able to transform protein into cell-building amino acids.

HOW TO BOOST PROTEIN METABOLISM IN MEAL PREPARATION

To help improve the usable amount of protein in foods, the Swedes follow these six basic steps:

1. Foods should be boiled, steamed or poached, and not fried.
2. Serving portions should be equal to the demands of the day's activities. Do not overload the system or put an excess strain upon the metabolism.
3. Serve fish at least three times a week for good ocean protein.
4. Serve fresh or sun-dried fruits instead of sugary sweets. The benefit here is that natural fruits contain good sources of protein metabolizing enzymes. Sugary sweets are excessively high in carbohydrates and displace protein, with the risk that cells and tissues may not then be adequately nourished or repaired. Sugar should be avoided.
5. Eat raw vegetables, or slightly steam and undercook those that have to be cooked. This offers a good supply of needed vitamins, minerals and enzymes that work upon protein for better metabolism of amino acids.
6. Serve whole grain breads. These contain good grain protein which is needed by cells and tissues for balanced nutrition. A balance with animal protein is important for enjoying the benefits of youthful health and vitality.

The hardy Swedish people maintain good health by emphasizing protein in balance with other essential nutrients. Since they are a hardy, healthy and seemingly youthful people, their programs should be utilized as closely as possible by others who seek to protein-ize their body's cells and tissues for a feeling of lasting youth.

THE SWISS "RAW PROTEIN" APPROACH TO
CELL-TISSUE REJUVENATION

The famed Bircher-Benner Clinic of Switzerland has long used "raw protein" as a means of boosting cell-tissue rejuvenation. It calls for using uncooked protein foods to give your DNA-RNA nucleus the highest biological pattern of amino acids that can be used for rebuilding with little extra effort. Many who have followed the easy, tasty "raw protein" approach in this famous Swiss healing sanatorium have reportedly undergone such a transformation that they felt the years roll away from their bodies and faces. Many tasted a return to the fountains of youth, thanks to the effectiveness of the "raw protein" cellular rejuvenation powers.

Why Swiss Youth Clinic Favors Raw Protein. This famous Swiss youth clinic has discovered a "secret" of the rejuvenation effect of raw protein. They have found that heating tends to destroy many enzymes present in the food. These enzymes are needed to work upon the protein factors in the food. Cooked foods have depleted enzymes, and the protein factors may not then be thoroughly metabolized. This suggests *fewer* amino acids available for cell-tissue rebuilding.

Raw protein foods contain a treasure of needed enzymes. These protein-substances work upon the protein in the foods to promote better transformation into amino acids. The enzymes then use these amino acids to create an increase of micro-electric rebuilding, better cellular respiration and more vigorous cellular metabolism. The enzymes promote amino acid rebuilding of cells so that the powers of immunity and resistance to aging are alerted. The Swiss Youth Clinic has found that raw protein foods are better able to promote the formation of red blood cells, and alert better respiration by nourishing the cells and tissues of the breathing apparatus. The raw protein has a stronger effect on nitrogen metabolism of the cell tissues, bringing about more youthful invigoration of the entire body. The Swiss Youth Clinic has long favored the increase in consumption of raw protein foods.

Raw Foods Offer Energetic Metabolism of Protein. For more effective cell-tissue replication, the body needs a more energetic metabolic system. The ability to absorb the components (amino acids) of proteins through the cell-tissue walls is influenced by energetic metabolism. Raw foods offer an abundance of enzymes that can more effectively transform proteins into usable amino acids that work to replenish the cells and tissues. This is the natural biological rhythm of rejuvenation. Raw foods promote youthful metabolism, and offer hope for lasting youthfulness through better amino acid nourishment.

THREE RAW FOODS USED AS PROTEIN HEALERS BY THE SWISS YOUTH CLINIC

The Bircher-Benner Clinic uses no animal products in their youth restoration programs. But they do recognize that protein is essential in order to supply the material that builds up the body cells and tissues, and that it protects against breakdown or "aging" by wear and tear. The Swiss healers suggest using these three raw foods as effective cell-tissue rejuvenators:

1. *Cheese.* Soft cheeses (such as cream cheese, cottage cheese) and others which are not too rich, can be more easily digested than hard or strong-flavored cheeses (which should be avoided, according to the healers). Soft cheeses contain protein in a form that is more readily assimilated.

2. *Eggs.* The Swiss Youth Clinic insists upon absolutely fresh eggs. They reportedly can be digested raw. Patients at the Clinic are given lightly beaten raw eggs, mixed with fruit juice and a little brown sugar. Raw egg yolk is often prescribed since it contains complete protein, which can stimulate bile in the case of a sluggish gall bladder. Egg white is pure protein, too, and can be eaten regularly. The Swiss Youth Clinic suggests a raw egg twice a week for good protein healing.

3. *Milk.* The youth specialists prefer *unboiled* milk. It can be used as a creaming agent for cereal mixtures, puddings and junket. It can be used with a granola cereal, too. It has good binding properties which can boost cell-tissue replication. Soured milk or yogurt, as well as buttermilk, are other forms of good raw food protein. Patients seeking healing at the Swiss Youth Clinic are given these protein foods:

Sour Milk. Also known as yogurt, it is made of skimmed or unskimmed milk, buttermilk or milk beaten up with lemon juice. It appears to have a rich treasure of proteins needed for soothing the gastro-intestinal region.

Skimmed Milk. A high-protein food that is good for those on a fat-restricted program.

Milk Pudding. This is milk made with rennet, or also known as junket. Rennet or junket powder is sold at almost all health stores and most food markets.

Plant Milk. (Non-animal milk.) In an electric blender or mixer, combine 1½ tablespoons of peeled almonds, 1 teaspoon honey and 3/4 cup water. Mix until thoroughly blended. Strain, if necessary. Drink this high-protein "plant milk" as you would ordinary milk. It is a prime source of plant protein, and is suggested by Swiss Youth Clinic specialists to folks who seek (or require) very speedy cell-tissue replication and general health.

Simple Protein Rejuvenation Program. Folks seeking better health at the Bircher-Benner Clinic are put on a simple protein rejuvenation program as follows:

1. Each and every meal must begin with a raw food or raw salad. The benefit here is to stimulate the digestive juices so that enzymes can better metabolize the protein into usable amino acids.

2. Daily, green leaves form part of the program. The benefit here is that plant protein helps restore a favorable acid-alkaline balance of the ashes after combustion of foods, more rapidly and more effectively. Green leaves supply a high biological value of protein together with a variety of vitamins, minerals and enzymes, which work harmoniously for better amino acid nourishment of the cells and tissues.

3. Whole grain breads and cereals are essential. The germ of the grain is a prime protein source which helps preserve the youthful freshness of the intestinal bacterial flora, and adds a feeling of comfort with any meal.

4. Seasoning should be *natural* and used in moderation. The benefit here is that harsh volatile seasonings such as salt or mustard can become corrosive in the system and interfere with protein and body metabolism. They suggest the use of simple herbs and spices, such as ginger, paprika, or curry used in moderation.

5. Meat is not served in the clinic since their specialists feel that it increases the amount of toxins and uric acid in the system. They suggest that meat be an *accessory* or an *exception* in the food program.

General Guidelines: All foods should be rich in natural fluids, in good fibrous bulk, vitamins, minerals and enzymes. Emphasize whole grain foods and as many natural and non-processed foods as possible. The main meal of the day is at midday. Breakfast and supper are light meals in which raw, fresh fruits are emphasized. There should be no eating between meals.

These simple programs appear to create a healthful protein metabolism. Body cells and tissues are given a steady supply of healthful amino acids that do much to offer the look and feel of youth. Many have gone to the Bircher-Benner Clinic and heaped praises upon their youth restoration methods. These same methods can now be followed at home. Give your body the working materials out of which new cells and tissues can be built, and you hold the key to lasting youth, thanks to protein.

THE MEDITERRANEAN "PROTEIN TONIC" FOR SECOND YOUTH

Many of the exclusive and fashionable health resorts around the Mediterranean feature wholesome foods as part of their youth restorative programs. They emphasize one important discovery:

A healthy digestive tract can promote better protein metabolism and overall cell-tissue rejuvenation.

These exclusive resorts feature a special "Protein Tonic," which reportedly is able to promote a boost in sluggish digestive enzymes and thereby promote more vigorous metabolism of proteins into cell-tissue building amino acids. Here is how to make this simple beverage:

"Protein Tonic" for Second Youth:

Put 1 tablespoon of apple cider vinegar into 4 tablespoons of skim milk (or buttermilk), and then sip it slowly. Take just once a day.

Boosts Recovery of "Always Tired" Tourist. Mary D. was "always tired," and looked it, too. She had wrinkled skin, dull eyes, would frequently mumble her words, and always had to sit down, even after walking a short distance. Her problem was irregularity as well as a deficiency of stomach acids containing enzymes which would be able to break down protein into needed amino acids for cell rebuilding. When she came to this Mediterranean resort, she was immediately put on a wholesome food program. But more important, every day, after breakfast, she was to take the "Protein Tonic." It took five such mornings before she began to respond.

Her irregularity was ended. Her digestion perked up. She had better metabolism since an increase of enzymes now made it possible for her system to transform protein into those amino acids that boosted cell-tissue repair. Once she was "restored" from within, she looked and felt better. She was alert. Her skin smoothed out. Her eyes sparkled. She talked clearly and now could walk for miles with not the slightest need for sitting down. She had given protein the "charge" needed by stimulating her sluggish digestive system. The "Protein Tonic," which uses just *two* ingredients available everywhere at pennies per treatment, had worked miracles in again making her the youthful person she deserved to be!

THE RIVIERA "HI-VI-PRO POTION" WITH "GO POWER" VITALITY

Stephen E. had spent a tidy sum for an expensive "once in a lifetime" vacation on the exclusive French Riviera. He had hoped that he would be able to "bounce back" with health and stamina, as had many of his co-workers who raved about such a vacation at the costly Riviera resort. But he found that even though he enjoyed fresh air, soothing sunshine, and abundance of different foods, and all the recreation and entertainment he wanted, he was still exhausted and weary. He felt he had spent his time and money in vain. He noted a couple, Mr. and Mrs. V.M., much older than himself, who displayed double his vitality even though they might have been double his age! He noted they ate wholesomely, but they had one special beverage every single day. He prevailed upon Mr. and Mrs. V.M. for the recipe. They shared the "secret" with him. The V.M.'s said that the recipe had been given to them by an exclusive "young doctor" who treated many of the wealthy socialites from the world's glittering capitals. He said it offered a supercharging of needed protein for speedy cell-tissue rejuvenation, the key to lasting youth.

The "Hi-Vi-Pro Potion"—How to Make It:

Use a blender or liquefier. Put one-half cup of chicken or vegetable broth into the blender. Add 1 tablespoon of fresh watercress or parsley. Add a slice of coarsely chopped mild onion. Whirr the blender. When this mixture is pureed, add one tablespoon of fresh raw liver (calf, chicken or very young beef). Continue the whirring until the entire mixture has a smooth consistency. Then add one cup of hot chicken broth or some steamed tomato juice. Stir. Remove from blender. Drink promptly but slowly.

Benefits: The enzymes in the raw vegetables are able to metabolize the protein from the broth and the liver and transform it into usable amino acids. Since the liver is uncooked, it has a biological balance of protein that is *undisturbed* and *uncoagulated*, which appears to make it favorable for amino acid replenishment of body cells and tissues.

Feels Peppy, Looks in the Pink. A grateful Stephen E. tried the "Hi-Vi-Pro Potion" daily. He ate more wholesome foods, too. He soon experienced a feeling of "regeneration." His skin smoothed out. He was alert and eager. Life became beautiful. He looked so healthy, he was mistaken for one of the "beautiful people" who are frequent guests at the plush hotels around the French Riviera. He thanked his kind benefactors for this secret. He returned home, "in the pink." He continues taking his daily "Hi-Vi-Pro Potion," and enjoys all the goodness that his young life has to offer. Protein helped boost his rejuvenation feeling. He had to go to the expensive French Riviera to discover this simple secret that costs pennies per glass but offers a million dollars in the feeling of healthful youth!

THE YOUTH SECRET FROM THE ADRIATIC

Longevity amongst the peoples living in and around the Adriatic (sea arm of the Mediterranean between Italy and the Balkan peninsula) is legendary. These peoples not only attain advanced ages well beyond the century mark, but they *look* and *feel* young. They have hardy stamina, alert minds, rarely wear eyeglasses, hearing aids or other appliances of so-called old age. It is not uncommon for men in the 60's and 70's to father children. This indicates that they have a youth secret. Researchers have found that these peoples have a healthful food program, and the emphasis is upon protein. While many may be unlearned and uneducated, they have traditional secrets for youth extension and prolonged rejuvenation. They know that these secrets *work*. They are living proof of it. This is all the convincing that they need.

Protein Lunch in a Glass: Beat two fresh raw egg yolks into a glass of any unsweetened fruit juice. (Orange juice appears to work more effectively in promoting better enzyme-protein metabolism.) Then drink slowly. Long-living folks of the Adriatic will drink this for a breakfast or a lunch in a glass. You may enjoy a slice of whole grain bread, if you wish. *Benefits:* Enzymes in the fruit juice metabolize the pure protein in the egg yolks. They transform the protein into amino acids to be used for building and rebuilding the billions of body cells and tissues, and promote a feeling of lasting youth. Two or three such *Protein Lunches in a Glass* per week can work wonders for your body and your mind. It is a staple amongst long-living folks of the Adriatic. It may well be their most important secret of all.

The Europeans have always been envied for their robust health, hardy bodies and alert minds. They do not think of "old age," which is a common affliction of more affluent folks in America. Rather, they know how to use protein and good foods to give them lives that are long and healthy. They are living proof of the effectiveness of using proteins for lasting youth. Share their secrets and use protein for self-rejuvenation. Life will be longer—and more youthful.

SUMMARY

1. Europeans know that a high protein, moderate fat, moderate carbohydrate balance offers hope for a look and feel of lasting youth.

2. The Swedes emphasize protein as the star nutrient in meal planning. They are a hardy race despite the harsh climate and rough terrain. Protein has given them extended youth and vitality.

3. The famed Bircher-Benner Clinic has a "raw protein" program for cell-tissue rejuvenation. This Swiss Youth Clinic has helped many folks enjoy a new lease on youthful life, thanks to protein planning.

4. Three raw foods can be used as protein healers, based upon discoveries by the Swiss Youth Clinic.

5. Mary D. tried a "Protein Tonic" as used by folks of the Mediterranean. She cast off her "always tired" feeling, enjoyed better and more youthful digestion, had a clear skin, sparkling eyes, and unbridled vitality and energy.

6. Stephen E. enjoys "go power" vitality thanks to a "Hi-Vi-Pro Potion" he discovered being used on the Riviera.

7. Look and feel as youthfully alert as the centenarians of the Adriatic by eating wholesome foods and a *Protein Lunch in a Glass*, several times weekly. It offers a powerhouse of cell-tissue rejuvenating protein in a matter of moments.

16

THE PROTEIN YOUTH SECRET
FROM THE ORIENT

The soybean is a native of China. It is one of the oldest crops grown by man. The earliest record of the plant is contained in the Chinese compendium of health secrets, *Pen Tsao Gon Mu,* written by Emperor Sheng-Nung in 2383 B.C. He extolled the youth-building secrets of the soybean, praised its values in giving super health to the body and the mind. He dubbed it, "the meat without the bone." As such, the soybean has often been the sole source of food for the struggling peoples of the Orient. The Chinese nation exists today because of its use of the soybean as food. Here is a miracle youth food, perhaps *the only meatless food that can be an effective meat substitute.* It has enabled the peoples of Asia to survive in times of war, deprivation, crop failure, even famine for hundreds of centuries. The soybean survives and thrives under the most difficult of conditions. Its secret built-in power of *complete protein* has enabled the Orientals to survive and thrive in the midst of food famine and hardships. The soybean has done more than just nourish the peoples of the Orient. It has given them long and healthy lives. It possesses a form of protein that exerts a miracle of cell-tissue rebuilding.

Today, the soybean is gaining in popularity throughout the world. The soybean is being recognized as a miracle protein food that has a natural ability for promoting overall nutrition by providing the cellular membranes with the amino acids needed for regeneration. It is the most exciting youth secret to come out of the Orient in many years. Let us see what the Oriental "youth food" can do for your better health.

Complete Bean Protein. When you eat cooked soybeans in any form, you are feeding a *complete* protein to your billions of cells and tissues. Basically, you know there are 22 amino acids. Your body can manufacture 14. But there are 8 which your body *cannot* manufacture and must come from foods. These 8 are known as the "essential amino acids" since they are essential for cell-tissue feeding and must be supplied from food. *The soybean is one of the few meatless protein foods that contains all 8 of these "essential amino acids." They are: valine, lysine, threonine, leucine, isoleucine, tryptophan, phenylalanine and methionine.* The Oriental miracle youth food contains all of these 8 "essential amino acids" which are needed to synthesize substances which create cellular membrane walls, and to promote the youth restorative process of tissue rebuilding.

Biologically Complete Protein for Cell Youth. A Nature-created food with biologically complete protein is considered of the highest value in the process of tissue rebuilding. The soybean was created by Nature with biologically complete protein for its own self-sustenance. Now, all of these miracle properties can be used for rebuilding and sustaining the youth of the body's cells and tissues. The Orientals may not have possessed our modern laboratory techniques or knowledge of food science, but they knew that a diet emphasizing soybeans helped them survive, thrive and enjoy long lives. This gave the soybean the label of a "youth food" as well as being the "meat without the bone." *Today, we know that the biological value of a protein is dependent on the minimum quantity of any essential amino acid made available from it. This is true in the soybean. It is the key to cell-tissue respiration.* The soybean offers a biologically complete form of protein that reportedly offers *everything* in the way of *all* amino acids for cell-tissue rejuvenation.

How to Boost Working Powers of Soybean Protein. The Orientals noted that while the soybean could nourish and sustain them it could be improved if it was combined with another food. The reason here is that one amino acid in soy protein, *methionine,* is slightly limiting, or lower in potency than the others. In a balanced food program, it is an insignificant matter. But when you are seeking to boost cell-tissue replication, you want to *double* the effectiveness of soybean protein and correct the slight limiting of methionine. The Orientals have a tradition of combining soybean with another food:

Combine soy flour with wheat flour and use for baking whole grain foods. A mixture of 5 parts of soy flour combined with 95 parts of wheat flour gives you 19% more usable cell-tissue feeding protein than wheat flour alone. This mixture has double its cell-tissue growth promoting value.

The Chinese Secret of Super Soy Protein: The hardy peoples of the province of Mongolia were able to survive in the midst of wars, pestilence, famines, and virtually no agriculture by emphasizing a simple mixture for baked goods: *add 9 tablespoons of soy flour to 6 cups of wheat flour. Then use for any baked products.* The hardy Mongols could grow both soybeans and wheat, which are hardy products that can be raised with a minimum of effort. When pounded into flour, and then baked, this soy-wheat flour mixture imparted a feeling of vigor and strength. They considered it their "youth secret," and it was a staple food that was favored far and above anything else that might have been available.

Why Soy-Wheat Mixture Is Youthfully Healthy. Food scientists today know that this mixture is healthful because the weaker soy methionine is boosted by the stronger wheat methionine. This *amplifies* the amino acid potency. When metabolized, the 8 essential amino acids are now biologically balanced and promote better cell-tissue rebuilding. The membrane walls of the body's billions of tissues are basically made up of these amino acids, with an emphasis upon *methionine*. This important amino acid is boosted and enhanced in this combination and appears to promote better cell-tissue replication.

The Hawaiian Protein Food That Creates "Happy Health." Edna G. felt dispirited, even moody. There were "off days" when her blue moods made her so despondent that she would feel like weeping for no apparent reason. Often she would fly off the handle. She had such severe temperamental misbehavior that friends and family drifted away from her. Being alone made her more depressed. She disliked eating by herself and would go from one restaurant to the other in the hopes of finding some solace. Restaurant food was inadequate. Then she ate at a Hawaiian restaurant which featured a simple but high protein breakfast food. This was the start of the amazing transformation that changed depressed Edna G. from "gloom" to "happy health."

Hawaiian Breakfast for "Happy Health." The menu said that the ancient gods of Hawaii promised "happy health" and "young life forever" to those who ate this simple but effective food:

Sprinkle sesame seeds and one tablespoon of soybean flour over a whole grain cereal. Add sun-dried raisins and fruit slices. Add a cup of milk. Stir and then eat in the morning.

Edna G. felt that her spirits perked. Her thoughts became more coherent. Her gloom evaporated. She was cheerful again. She smiled. She laughed. She joked. She no longer lost her temper. She became so personable that her friends and family were constantly calling her for parties and social gatherings. Edna G. has changed her food program and has this easy Hawaiian Breakfast for "Happy Health" almost daily. It has made life a joyful thing again.

Oriental Secret of "Happy Health" Breakfast: Soy flour has more lysine than is required for the usual cell-tissue feeding of the body, especially of the brain. This means that soy flour should be *balanced* with another grain food that is low in this amino acid. Most cereals are *low* in lysine. When soy flour and whole grains are added, there is a better blending. Sesame seeds are very *high* in methionine but *low* in lysine. Therefore, when sesame seeds are combined with soy flour in whole grain cereals, the amino acids appear *balanced* in such a unique combination that they exert a beneficial cell-tissue replication of the billions of cells in the brain and other body parts. This helps create a better emotional *equilibrium* since *balanced amino acids help create balanced cell building.* When the brain cells, such as those of Edna G., are better nourished through balanced amino acid rebuilding, there is a reward in better emotional and mental health. The Hawaiian secret did, indeed, give her "happy health" thanks to balanced amino acid pattern rebuilding of the cells.

Simple Program: Supercharge your brain and body cells in the morning by preparing a simple whole grain cereal. Most health stores and supermarkets sell whole grain, unbleached cereals. Just add milk, a sprinkle of sesame seeds, one tablespoon of soybean flour and some fresh fruit. Stir and eat. This Oriental repast offers a feast of *balanced proteins* with *biologically balanced amino acids* for overall cell-tissue replication. You will find that it gives you a natural boost in the morning for prolonged efficiency of your body and mind. It is the Hawaiian secret for "happy health" that you can enjoy every day of your "youth forever" lifespan.

HOW THE JAPANESE USE SOYBEAN SPROUTS FOR VIRILE VITALITY

The Japanese are crowded in their cities and rural areas. Yet they are reportedly among the healthiest of the nations of Asia, and, perhaps, the world. To the Japanese, the soybean has always been revered as "secret of eternal youth" because it has served as the foundation for life and health in the absence of other foods.

The Japanese go one step further with their revered soybean by making it more potent for health building. The Japanese have long used *soybean sprouts* as a "secret" for prolonged and extended virile vitality. They have discovered that when soybeans are sprouted, the natural amounts of vitamins, minerals, enzymes and proteins appear to become *magnified and increased.* Sprouting boosts the complete amino acid pattern and makes it even more compatible for better cell-tissue replication. When soybeans are sprouted, their vitamin content appears to enhance the absorptive values of their amino acids. It is this *harmony* that helps promote a more *biological harmony of tissue rebuilding.* The Japanese have thrived and survived during times of near starvation on the tradition of eating high-protein soybean sprouts. *Many modern Japanese affirm that soybean sprouts offer the highest biological value of complete amino acid patterns of any food.* They are living proof of the effectiveness of the life-sustaining powers of protein-ized soybean sprouts.

How to Grow Soybean Sprouts the Japanese Way. Here are the simple step-by-step directions for growing soybean sprouts in your kitchen:

1. Put two tablespoons of raw soybeans in a quart jar. Half fill the jar with tepid water. Cover and let stand at room temperature overnight.

2. Next morning, cover the jar with a double layer of cheese cloth, nylon net or a screened lid.

3. Now pour off the soak water (without removing the screened lid) into another jar. (*TIP:* The healthy Japanese do *not* discard this high protein water. They mix with fruit or vegetable juices; or they use it as a soup stock. Use it for cooking brown rice. It's a precious treasure of vitamins, minerals, enzymes and proteins that work harmoniously to supercharge the billions of body tissues with electrifying vitality and renewed health.)

4. Next, rinse the soybeans with tepid water. After rinsing, tilt the jar slightly to permit remaining moisture to escape; but the soybeans should still have a source of oxygen.

5. Put the jar of soybeans in a dark place. (*TIP:* Modern Japanese prefer beneath the sink since the atmosphere is comfortably warmed by the hot water pipe.) Otherwise, rest the jar on the sink top in a tilted position. But the soybeans should be kept dark.

6. Repeat this rinsing process about four times a day.

7. By the end of your second day, you'll see your soybeans

starting to sprout. By the morning of the third day, they will continue to grow. You may refrigerate them. The soybeans will keep on sprouting, magnifying in protein content while being refrigerated. Plan on using them within a few days.

HOW TO ENJOY PROTEIN SOY SPROUTS

The Japanese have devised these tasty ways to feed protein soy sprouts to their body cells. Ingenious as they are, they are very clever in making simple foods become a feast with these delicious techniques:

Salads

Serve raw sprouts in a raw vegetable salad. Season with a bit of vegetable oil. A sprinkle of lemon juice adds a piquant taste.

Hot Protein Vegetable

Pan broil soybean sprouts lightly. The Japanese use very little oil since the soybean has its own built-in reservoir. Use as a hot vegetable.

Soybean Sprout Plate

Pan broil some sliced onion in a little bit of oil. Add the soybean sprouts and a few tablespoons of water. Cook less than 10 minutes. Makes a delicious and super-high protein salad.

Egg Foo Yung

This Chinese specialty is featured in many Japanese restaurants. Just add chopped sprouts with soy sauce to omelets or scrambled eggs. Makes a terrific high-protein breakfast with exceptionally good amino acid balance. It is the staple food throughout the Orient.

Soups

Any soup can be boosted many times in protein content by just a handful of sprouts which you drop into the soup before serving.

Snacks, Munching

Keep a jar of soybean sprouts in your refrigerator and use for delicious snacks and munching. The Japanese use them as "pickups" when they need a "lift."

The amino acids of the soybean sprouts go to speedy work in revitalizing and rebuilding cells and tissues to promote a healthful boost to energy.

Protein Sauce

Any sauce becomes a powerhouse of protein when you add soybean sprouts.

Main Dishes

Mix soybean sprouts to chopped meat for high-protein hamburgers. Mix with chopped chicken or any diced foods that can be baked, broiled or, in the case of plant foods, that can be eaten raw. The soybean sprouts have a rich, chewy texture and can add zest to any food.

Protein Dessert

Mix one cup soybean sprouts with 1 cup of any nuts or seeds; then add 1 cup sun-dried, unsulphured raisins and 1 tablespoon honey. Run this mixture through a food chopper. Add a bit of unsweetened coconut flakes. Mix well. Shape into balls. Eat as a powerhouse of protein for a healthful dessert. This easy-to-make, uncooked delicacy is featured in many Oriental restaurants. In a few moments, you can make it yourself and enjoy a feeling of "lifetime youth."

HOW TO ENJOY HIGH-PROTEIN SOYBEANS

The Orientals follow this easy cooking program:

1. Wash soybeans and put them in a kettle partially filled with water. Let soak overnight. Since soybeans will expand, you should use 1 cup of soybeans with 3 cups of water, as do the Orientals.

2. Next morning, bring to a boil in the same water. Reduce heat and cool slowly. Let simmer for 2 hours or until the beans are a light tan. Sample some of the beans. If they are chewy, then they can be eaten.

3. Eat a plate of soybeans, together with a large raw vegetable salad seasoned with lemon juice and vegetable oil. This is a delicious, simple, but super-high protein meal. For millions of Orientals, this easy-to-make meal served as their entire sustenance. It was a miracle protein food that gave them life and health.

SOY GRITS: THE HONG KONG SECRET OF SUPER HEALTH

Alma F., a textile buyer, was always feeling tired. She walked with a tired or stooped and shuffling gait. The streets seemed longer and longer. Every staircase seemed higher and higher. She was constantly short-winded. Even if she ate wholesomely, she still felt under par. Then one spring Alma F. went on a business trip through the textile manufacturing district of Hong Kong. Here she noticed that very elderly Orientals, approaching the century mark, displayed amazing vitality. Women in their 80's and 90's, who lived on houseboats, could shoulder heavy bundles and shuffle through the narrow, winding streets, with amazing agility. What was their secret? She looked into their kitchens and was told by a wizened, smiling octogenarian lady that "soy grits" was the "miracle youth food" of Hong Kong and the Orient itself. She told Alma F. how to use them and gave her a big crock of soy grits.

Straightens Up, Feels Energized, Emotionally Alert. Alma F. began to use soy grits in her meal planning. Within four days, she was able to straighten up. Her body and mind felt supercharged with energetic alertness. She could walk for long miles without feeling unnecessarily tired. She could breathe healthfully. She felt rejuvenated. Now she uses this Hong Kong secret of super health regularly.

How to Make Soy Grits

Chop soaked soybeans into smaller pieces so they will cook faster. Mix these soy grits with any meat loaf, vegetable loaf, or any chopped salad. Mix with cereals. Mix with whole grain unbleached flours to boost protein content of baked goods. Mix in scrambled eggs and use as part of stuffing. Use in casseroles.

Protein Benefits: Soy grits have a treasure of nutrients, particularly protein. *One tablespoon of soy grits will add 6 grams of cell-rebuilding protein to any dish.* The hardy, long-living people of Hong Kong use soy grits for almost all of their meals. Many are so impoverished that they can scarcely or rarely afford to buy meat, fish, eggs or dairy foods which would give them protein. But they *always* buy soybeans and *always* use soy grits in the most humble of meals. They know, through thousands of years of tradition, that soy grits offer the "life's blood" of health through our modern knowledge of protein.

For the Orientals, the soybean is the staff of youthful life!

THE MEATLESS "PROTEIN MILK" THAT PROVIDES YOUTHFUL MOISTURE TO THE BODY'S THIRSTY CELLS AND TISSUES

The Orientals appear to have unlimited energy and stamina, despite famine, difficulties or holocaust. Much has been traced to their use of the soybean in all forms. But one use, recently discovered, appears to provide a miracle of youthful vitality to their bodies and minds.

Soy Milk Protein Moisturizes Cells. The Oriental makes milk out of the soybean. Drinking this high protein milk sends a supply of youthful moisture to be readily absorbed by the body's thirsty cells and tissues. Beneath the surface of the skin, under a microscope, can be seen a honeycomb web network of deep "wells" or "reservoirs" that are "thirsty" in that they require liquid in order to remain moist and youthful. Soy milk protein appears to offer this type of moisture to keep the "wells" properly filled and promote a look and feel of youthful health. The Oriental has tapped the secret of perpetual or lasting youth by making soy milk a staple food.

How Orientals Make Soy Milk: In a kettle, place 1 cup of soybeans. Add 3 cups of water. Let soak up to 48 hours. Next, pour off the water and use for a soup or casserole. Grind up the soybeans through a meat chopper or put in a blender and chop until they are *very fine.* Place in another kettle. Add 3 to 5 cups of fresh water. Boil for 30 minutes. Strain. Add a little honey, if desired. You now have *soy milk,* the meatless "protein milk" that is a powerhouse of "liquid nourishment" for your amino acid-requiring cells and tissues.

For those who prefer a meatless milk, this easy-to-make soybean milk is the answer. It is a rich treasure of *liquefied amino acids* that become absorbed by the "parched" cells and tissues and create youthful moisture, the key to lasting youth. The Orientals know of this secret and use it to perpetuate themselves in the absence of dairy milk. Now, soybean milk can be enjoyed by everyone for better cell-tissue health and better vitality.

Orientals Thrive on Soy Protein. The peoples of the Orient have looked upon the soybean as their "staff of life." Our modern food scientists have learned that soy sprouts are superior in protein. The sprouting process converts the protein into a higher form, a more biologically balanced form that is more digestible and better assimilated by the body's cells, tissues and organs. The Orientals enjoy long, vitalic and virile lives, thanks to their protein-ized health because of soybeans in all forms.

The soy is the greatest protein bean of them all. It is the food that gives long life to the Oriental. It can help you, too. When you use soybeans to protein-ize your body, you are using the oriental secret of longer life. So did one Japanese lady, Mrs. Mito Umeda, believed to be the oldest woman in Japan. She is 110 years old. She stated[1], "I eat any kind of food, but I never miss taking soybean flour as well as fruit once a day." Protein is, indeed, the "staff of life."

IMPORTANT HIGHLIGHTS

1. The Orientals have long looked to the soybean as a protein food that gives them long life and health.

2. The soybean is one of the few meatless foods that offers *complete* protein and all 8 essential amino acids which your body cannot make but must take from foods eaten.

3. Combine soy flour with whole wheat flour for more usable soybean protein. It is the Chinese secret for lasting youth.

4. A simple, tasty Hawaiian Protein Food helped Edna G. wash away emotional gloom and enjoy "happy health."

5. The vitalic-virile Japanese thrive on super-protein soybean sprouts. You can grow this type of protein in your own kitchen.

6. Soy grits, as discovered in Hong Kong by Alma F., gave her a feeling of super health and freedom from premature aging and tiredness.

7. For an Oriental meatless "protein milk," make soy milk and quench the thirst of your millions of body cells and tissues. It's the tasty way to *drink your protein,* the "staff of life."

---

[1]*Associated Press* release, March 25, 1973.

17

HOW TO COMBINE TYPES
OF PROTEINS FOR
DOUBLE-ACTION REJUVENATION

Give your cells and tissues a rich protein harvest of *high biological value* and the DNA-RNA components are able to revive and re-generate with double-action rejuvenation. The membrane walls of the billions of body cells can respond with better healing and better replication when nourished with protein that is *balanced* in a good combination form.

Complete + Incomplete Proteins = Double Action Cellular Rejuvenation. By combining different forms of animal and/or plant proteins, you fuse together the complete and incomplete proteins to create a combination-balance that offers a form of double-action cellular rejuvenation.

The youth-building quality of a protein is frequently indicated by its biological value. Biological value is a term used to express the percentage of absorbed nitrogen which is retained in the body and used for cellular maintenance, growth and rejuvenation.

Animal Proteins Are Complete: Animal proteins, such as those from meat, poultry, eggs, milk and milk products, and fish are considered to have *complete* proteins and have a high biological value.

Plant Proteins Are Partially Complete. Plant proteins, such as peas, beans, nuts, cereals, grains, lentils, soybeans, seeds, fruits and vegetables are considered to have *partially complete* proteins because they are lower in one or more of the essential amino acids.

Your Body Cells Require Protein Combinations for Better Repair, Regeneration, Rejuvenation. When your body receives a *combination* of complete and partially complete proteins, it is able to convert them into strong amino acids that then enter into cellular repair,

189

regeneration and rejuvenation. The potency of the absorbed nitrogen is boosted and made more vigorous when proteins are combined, and this helps build better and stronger cells and tissues.

Plant foods contain combinations of protein which together can make for a better protein quality than each taken singly.

Simple Combination: For better cellular replication, try mutual supplementation of proteins. For example, you require slightly more than a 3:1 ratio of the amino acids, lysine to tryptophan, a ratio which a combination of milk, eggs and meat can provide. Another combination calls for eating both milk and cereals together because milk is low in methionine and cereals are low in lysine. When you combine milk with whole grain cereals, the combination of lysine by milk and methionine by cereals improves the utilization of both proteins. By combining different protein foods, you can give your body a high biological value of good cell-feeding nitrogen, and thereby give your billions of tissues the needed working materials for regeneration.

THE NET PROTEIN UTILIZATION (NPU) WAY TO CELLULAR REJUVENATION

Your body's protein consists of some 22 amino acids. Most of these can be made by your body. But eight are called *essential amino acids* because your body cannot make them. Yet, they must be available to your body at the *same time* and in the *right proportions* for your cells to be adequately nourished with high biological values.

If this does not occur, then your cells are only *partially* nourished. This is like trying to run your car with some of the cylinders blocked.

Many plant proteins are incomplete or weak in some amino acids. Therefore, they must be balanced with other protein sources at the same meal in order to permit your body to absorb their protein fully. This is called the *Net Protein Utilization,* or NPU—the key or secret for better cellular rejuvenation. When your cells receive a *combination* of protein foods, they are fed the NPU method which creates better membrane rejuvenation and healing.

FIVE MIRACLE PROTEIN COMBINATIONS FOR NPU CELL HEALING

Here is a simple guide to help you learn how to combine proteins for better nitrogen formation and *net protein utilization,* or NPU cellular revitalization.

1. *Combine Grains with Animal Proteins.* Most whole grains and cereals contain good protein, but wheat has varying amounts of amino acids. Gluten flour is high in protein. French breads are usually made with "hard" or higher protein flour. But other types of grains and flours may be *lower* in protein. So combine whole grain breads with *smaller amounts* of milk, cheese, meat, fish and poultry.

Suggestion: Rice cooked with milk, or whole wheat eaten together with a handful of nuts, will balance each other's amino acid pattern. This makes for a very efficient protein utilization by the body's cells and tissues. Rice proteins are somewhat partially incomplete, as compared with meat, fish or cheese, so you would need more of them to satisfy your cells and tissues.

2. *Combine Beans, Peas, Lentils, Soybeans.* While most beans, peas and lentils do have a very high proportion of protein, they are only partially complete in certain amino acids such as tryptophan. Therefore, you can correct this by combining different types of these foods with other foods, such as rice, whole wheat, milk, peanuts or sesame seeds. This results in a super-protein balance of high biological value and better NPU absorption.

Suggestion: Prepare Boston baked beans with nutted brown bread and butter for a good NPU balance. Or make a chili meal or any dish of beans, peas and lentils together with natural brown rice and cheese. This creates a double-action potency of amino acids that work vigorously and effectively for cellular rejuvenation.

3. *Combine Seeds, Nuts, Grains with Some Animal Foods.* This gives you an assortment of different types of amino acids that blend with one another for better balance. A peanut butter sandwich on whole grain bread, together with a glass of milk, can give you a good amino acid pattern for satisfactory NPU absorption. Feature a dish of any type of combined seeds, nuts and whole grains with cheese or milk.

Suggestion: Make a main dish of cooked beans with chopped nuts and a sprinkle of assorted seeds. Add a scoop of cottage cheese or yogurt. You now have good NPU balance of usable amino acids. You can also make a high-protein "burger" by chopping nuts and mixing with grains and cooked beans, and then shaping into a patty. Use chopped vegetables and an egg to bind this "burger" for balanced amino acids that are complete and powerful for cell-rejuvenation.

4. *Combine Fruits and Vegetables with Some Animal Proteins.* Plant foods such as fruits and vegetables give you partially complete

or "penny-bank" amino acids in smaller proportions. While not complete, they are helpful and should be eaten every day to provide the balance needed by your cells.

Suggestion: You can boost the amino acid potency of plant foods by combining them with cheeses, nuts, rice, chopped hard-cooked eggs, or small amounts of fish, poultry or lean meat. Add flavor with a sprinkle of lemon juice or tomato juice.

5. *Combine Dairy Products with Eggs.* The protein of eggs is almost fully utilized by the body. It is noted that both eggs and milk (and all the cheeses) are better utilized than beef. Dairy products have a good NPU ability. Dairy products have a complete amino acid profile and are especially vigorous in specific amino acids. This makes them highly desirable when combined with eggs as well as whole grain breads and cereals.

Suggestion: For a good morning cell-feeding plan, try scrambled eggs with cheese and a slice of whole grain bread with a pat of butter. Or try a sandwich of hardboiled egg slices with cheese on whole grain bread. It's a tasty, filling way to feed high biological values of protein to your cells and tissues.

Everyday foods are prime sources of the cell-feeding amino acids. Since each food has varying amounts, potencies and percentages of these amino acids, it is healthful to *combine* a variety of foods to create good nitrogen formation and better *net protein utilization* (NPU), the key to cellular revitalization. Combining these foods is the key to better cellular replication.

20 TYPES OF PROTEIN COMBINATIONS FOR DOUBLE-ACTION CELLULAR REJUVENATION

When you *blend* several everyday foods together, the biological pattern of the amino acid content is boosted. It is comparable to the strong man who helps the weak one. Together, they can accomplish efforts that, singly, might be difficult. So it is with *protein combinations.* A weaker amino acid of one food becomes *naturally enriched* and *naturally fortified* when it is eaten together with a stronger amino acid of another food. It creates a biological balance. It is a way to give your cells a double-action rejuvenation with stronger *net protein utilization.* Here is a listing of 20 types of protein combinations that offer gourmet taste and gourmet health building qualities. *No food is protein weak when it is combined with a protein strong food!* Here is this listing of good NPU cell-building food combinations:

1. *Combine natural brown rice with beans or peas.* You boost the usable amount of protein for better biological body balance.

2. *Combine natural brown rice with soybean foods.* The rice protein is energized by the superior soybean protein. You may use cooked soybeans, or any soy flour product for this good balance.

3. *Combine natural brown rice with wheat and soy foods.* When eaten in this combination, you amplify the weak protein of the wheat because the strong protein of the brown rice appears to boost its utilization; the soy protein offers a complementary proportion of more complete proteins for a good balance.

4. *Combine natural brown rice with Brewer's yeast.* The amino acids of both of these foods offer better nitrogen utilization when eaten in this simple but effective combination. Just add 2 to 4 tablespoons of Brewer's yeast for each cup of natural brown rice. Use milk instead of water for even better protein utilization. It creates good amino acid balance and potent protein.

5. *Combine natural brown rice with sesame seeds.* The amino acid pattern of sesame seeds is of a high biological value. While some of the amino acids are weak, they are enhanced when you combine this food with natural brown rice. This combination enhances the weak amino acids and gives them "go power" for better cellular replication.

6. *Combine natural brown rice with milk.* The amino acid pattern of the natural brown rice appears to complement that of milk. There is better cell utilization of amino acids when nourished with this combination. You may use any milk product such as buttermilk, yogurt, cheeses or skim milk.

7. *Combine whole grain wheat foods with milk or cheese.* A high biological protein pattern appears to give a boost to a lower or partially complete protein pattern. The high protein of the dairy food adds emphasis to the partially complete protein of the whole grain wheat foods. You can give yourself good NPU cell replication by having a bowl of whole grain cereal with milk. Also, select whole grain bread products and use them when making cheese sandwiches. Sprinkle wheat germ over cereals and milk, and then add fruits for a triple effect; that is, the enzymes of the raw fruits will boost the protein of the wheat and the milk to create favorable cell regeneration.

8. *Combine whole wheat foods with beans.* A partially complete amino acid pattern of bulgar wheat, of buckwheat, or of any product made with whole wheat becomes completed and better assimilated

by the cells when eaten together with partially complete beans of any kind. A bean *mixture* will further enhance the protein metabolic function of the combination. Make a bean casserole with bulgar wheat and an assortment of fruits and vegetables for good biological nourishment.

9. *Combine whole wheat foods with soy foods.* Whole wheat proteins are valuable but they are somewhat weak, and they need the boosting of soy food proteins. For example, whole wheat is low in lysine, but soy food is high in this amino acid. Yet, soy food is lower in methionine and cystine, but wheat is high in these amino acids. When *combined,* these amino acids appear to enhance one another and your cells are given a better balance of nutrients for overall utilization. Mix whole wheat and soy flours in baked goods for good biological balance of protein.

10. *Combine whole wheat, soy and sesame seeds.* The amino acid patterns of these three foods complement one another in a combination. There is an increase in the utilization of their amino acids and a mutual supplementation for the cells and tissues. Mix whole wheat with soy flours, and use sesame seeds and sesame flour, too, for good protein balanced foods.

11. *Combine natural cornmeal and bean foods.* Natural cornmeal, or that which has not been treated by chemicals, has a good protein supply. But it is low in such amino acids as lysine, methionine and cystine. But bean foods are high in these amino acids. When you combine the two foods, you complement the proteins and bolster the weaknesses. You then have a high biological value of protein that is healthfully vigorous for the cells and tissues. A casserole or stew made with cornmeal and any type of beans is one that is of high biological protein value and benefit.

12. *Combine natural cornmeal, soy foods and dairy products.* Boost the lower methionine and cystine amino acids of the soy foods with natural cornmeal and dairy products. It creates better balance. You may use any dairy food, such as milk, cheese or yogurt for its amino acids. Skim milk, too.

13. *Combine milk and beans.* Milk contains casein which represents about 81 percent of its total protein content. This protein enhances the lower amount of amino acids in most beans. When you have a simple bean dish with a cup of milk (or any dairy product), you boost the mutual acid patterns and create a high biological value.

14. *Combine beans and sesame seeds.* These two plant protein

foods have varying amounts of amino acid patterns. To make up for any weaknesses, combine both of them for a complementary effect in cell-tissue nourishment.

15. *Combine soy, wheat, rice and peanut foods.* Here are four high-protein foods which contain different amino acid patterns. There are slight weaknesses in each of these plant foods. But when you combine *all four* in any dish, then the weaknesses are bolstered. The amino acids *overlap* so that the strong ones will bolster the weak ones. The benefit is you then have a high biological pattern with stronger net protein utilization (NPU) for cells and tissues.

16. *Combine soy, peanuts, sesame seed foods.* Here are high protein foods, but each has one or two lower amino acids. But when combined, these weaknesses are overcome. The result is a super-charged protein food. Double-action cell rebuilding and a good biological pattern are offered in this simple combination.

17. *Combine peanuts and sunflower seeds.* These are two top level protein foods, but slight amino acid weaknesses can be corrected when you combine them at a single meal. You can have one-half cup of shelled peanuts with one-half cup of shelled sunflower seeds as a good protein booster.

18. *Combine peanuts, wheat products with milk foods.* You will give a special boost to weak amino acids by combining peanuts and wheat products with milk, which has an almost complete protein pattern. The weak amino acid in milk is then boosted by the strong peanut and wheat proteins. You can mix peanut and wheat flour with milk for baking. Or just have a whole wheat bread sandwich with peanut butter, a slice of cheese, and a glass of milk. This is the delicious way to give your body a balanced biological balance of complete protein.

19. *Combine milk with sesame seed foods.* The weak amino acids of milk are boosted by the strong amino acids of sesame seed foods. You can use sesame seed flour (sold at many health stores and specialty food shops) in a baking dish with milk for a good biological amino acid pattern. Or just sprinkle sesame seeds (sold at health stores and specialty food shops) over your whole grain cereal, to which milk and fruit have been added.

20. *Combine milk with potatoes.* The weaker amino acids of milk become amplified by the stronger amino acids of the potatoes. If you'll mash potatoes together with milk, and then sprinkle with desired herbal seasonings, you'll have a tasty and healthful balance of good protein.

These 20 combinations are aimed at getting the most out of the protein in foods. When you plan your meals around such easy and delicious combinations, your cells will be receiving exceptionally good amino acid patterns needed for double-action replication.

SIMPLE PROTEIN PLANNING CORRECTS PREMATURE AGING

Dennis J. was one of those types who ate "everything and anything." He looked and felt it, too! His hands trembled. He had aging spots on the backs of his hands. His face had unsightly blotches. He felt weak and tired. Walking a short distance tired him. He could hardly do ordinary gardening since a slight bending over gave him a wrenching pain. A kindly neighbor, who was double his age, but had double his energy, told him how to use simple protein planning for a natural protection against so-called aging.

Easy Program Promotes Well-Being. Dennis J. started to combine different proteins. He would have a lentil-rice loaf for lunch. Or he would have a corn-lima bean and succotash dish. His breads were made of a combination bean and grain flour. He had whole grain bread sandwiches of peanut butter and cheese, with a glass of skim milk; or he would have a yogurt with wheat germ and fresh fruit slices. He would use nuts and seeds as snacks in place of the sugary confections he habitually devoured. He would have more fresh green salads. He would often have a meal of mashed potatoes with milk, a raw vegetable dish, whole grain bread slices, fresh fruit and a cup of herb tea or Postum.

Feels Energetic, Looks Cheerful. Dennis J. discovered that when he followed as many of the protein combinations as possible, he began to feel energetic. He started to look and feel cheerful again. He had amazing vigor. He could now walk for miles and then do work around the house with amazing stamina. His hands no longer trembled. The aging spots vanished. He looked "in the pink" thanks to protein planning. Now he considers protein the "star" in his meal planning. It has made him feel young and glad all over.

Benefit of Protein Planning: Cells and tissues need a balanced amino acid pattern created by additional nitrogen. This is a characteristic and relatively constant ingredient in protein. Sufficient nitrogen made available for the cells and tissues means a regular supply of those ingredients that enter into the membranous walls and promote internal rebuilding and regeneration. When you use combined foods in your eating program, you have sufficient nitrogen to maintain body homeostasis and cellular replication. It's the delicious way to feed yourself lasting youth!

The human metabolic system cannot store protein in the manner as it does fats and starches. All unusable protein is excreted. While you may exist comfortably on a minimum intake of most nutrients, you need an adequate intake of protein, which is the nutrient that prompts the utilization of all other nutrients to create the substances for cell-tissue replication.

The effectiveness of protein is determined by the amount of protein retained by the body. You may be eating good protein foods, but they may not be stored in the body, or weaker amino acids are not being properly utilized. This means that your cells and tissues, and your body, are not receiving all the nourishment required. By emphasizing a *balanced combination of variety*, you will be supplying your body with the required amino acids to create a favorable *net protein utilization* (NPU), the key to better and more lasting cell-tissue rejuvenation. It offers hope for lasting youth.

IN REVIEW

1. Combine complete animal proteins with partially complete plant proteins for better *net protein utilization* (NPU). It promotes better cell-tissue replication.

2. There are five tasty miracle protein combinations that boost NPU cell healing.

3. There are 20 types of protein combinations of everyday foods that help promote double-action cellular rejuvenation.

4. Dennis J. combined foods in a simple but effective way. His body was supercharged with vitality. His "shaky" hands steadied. The "aging spots" vanished. Now he worked with the vigor of a youngster, thanks to simple protein planning for cellular recharging.

18

THE AMAZING
REJUVENATION POWER OF
SEED AND NUT PROTEIN

The protein of seeds and nuts possesses an amazing ability to reconstruct the billions of body cells and tissues. This ability is unique, and unlike other protein foods, *seeds and nuts can synthesize their own protein for their own regeneration and sustenance.* They are self-sufficient and self-rejuvenating because of their ability to create the miracle of cell-building protein. It is a miracle of Nature that is comparable to the Fountain of Youth, since seeds and nuts can continue their own self-perpetuation for lasting life by creating their biological balance of protein. When you eat seeds and nuts, you are giving your body the same biological pattern of amino acids that has helped give the plant its self-perpetuating youthful life.

The Cell-Building Powers of Seeds and Nuts. The biological balance of amino acids in seeds and nuts is harmonious and in a pattern that favors superior cell rebuilding. This pattern is especially favorable to the cells. Basically, the nucleic acids DNA and RNA are the cellular components which control the ability of the body to keep reproducing and rebuilding its *inherited* or *genetic* (inborn) patterns.

Seeds and nuts have a balanced protein that acts to nourish the strands of DNA and RNA within the cell nucleus. The seed-nut protein appears to protect against the oxidative reaction wherein cross-linkage or malformation of cells may cause breakdown and the aging process. Seed-nut protein offers a form of cellular reparation by nourishing the DNA-RNA strands within the nucleus and creating a "shield" to protect against malformation. Just as the protein has served to keep the seed and nut in good and lasting health, so can it help to nourish your own cells and the essential DNA-RNA components that extend and perpetuate tissue regeneration.

HOW PREMATURE AGING CAN BE REVERSED
WITH SEED-NUT PROTEIN

Briefly, the body produces free radicals (irregular electrons) which adversely react with the body's spiralling protein factory, or DNA-RNA. Successive reactions between the radicals produce a breakdown of the DNA which leads to peculiar or deformed cell creation. This causes the symptoms of tissue breakdown and aging. Food technologists and youth scientists suggest that if the body can be *protected* against the malformation of irregular cells, then the premature aging tide could be reversed. The protection should come from a food that has the source of lasting youth within itself; namely, a form of protein that can sustain the food without any other requirement. If the food can remain "forever young" with its own protein, then human beings could have hope for lasting youth if given the same protein. Today, we recognize *seeds and nuts as having that special protein which can promote lasting youth by protection against irregular cell formation.*

The Miracle Protein Power of Seeds and Nuts: Packed in a Nature-made, waterproof and air-tight shell, the nut meat is a biological balance of self-sustaining protein. It is clean and wholesome. Hermetically sealed by Nature, the nut does not become contaminated and spoiled such as meat, eggs, dairy or just about any other food would. Nuts are free from waste products, are aseptic and do not readily decay, either in the body or outside of it. They are not infested with harmful additives, as would be the case with most chemicalized foods. Seeds and nuts have very high biological protein which keeps them in a state of lasting youth. This balanced amino acid pattern makes seeds and nuts a treasure of the substances needed to offer cellular replication and a feeling of lasting youth for those who eat such healthful foods.

Rich Treasure of Vitamins, Minerals, Enzymes, Essential Fatty Acids. Seed and nut trees or plants strike their roots deep into the soil where they take up needed vitamins, minerals, enzymes, essential fatty acids and protein. This creates a treasure of *combined* nutrients offering an all-purpose benefit to the cells and tissues. Seeds and nuts contain almost all known nutrients that are used by protein for the rebuilding of its own core. This same benefit can be enjoyed by your body when you enjoy seeds and nuts in daily food programs.

High Biological Value. Seed and nut proteins are considered to be of high biological value. The body draws upon such proteins for its supply of tissue-building material. Seed and nut proteins are metabolized into a high biological form of amino acids which help to adequately maintain youthful life and growth.

Purest Protein Known to Man. Organically raised seeds and nuts are believed to have the purest protein known to man. While animal foods, and even plant foods, contain complete or pure protein, it is difficult to obtain meat produced from cattle that has not been injected with chemicals to fatten the cattle for the market. It is difficult to obtain meats from healthful feeding sources. Grains and grass are polluted because of atomic fallout in the fields. Spray residues are ingested by cattle, and are imbedded into their tissues. Plants grown on such soils are also chemically infiltrated. This alters the protein pattern. It can create an altering of the body's cells and tissues.

But seed and nut protein can be considered *pure* since the shell or covering of the food is thick and offers a form of protection that has no comparison. To feed your DNA-RNA cell components a form of *pure* protein and to protect against the formation of free radicals or cross-linkage which can be traced to chemical ingestion, it is healthful to seek out the *pure* protein of seeds and nuts.

SEED AND NUT PROTEIN: THE MIRACLE OF LASTING CELLULAR YOUTH

The *pure* protein offers a *thorough* method of cellular replication. The seed and nut proteins are not used by the body as such. They are split into smaller components during digestion and then are absorbed into the bloodstream.

Protein Pool in Reserve. The metabolism of proteins provides a continuing source of amino acids which forms a "pool" for use, in reserve, as needed. It is from this mixture of molecules that the body selects the kind and number of amino acids required to build new cells, repair existing ones, and create such body hormones and enzymes that prolong and extend the prime of life. Seed and nut protein is self-sustaining or self-fulfilling. It can keep the seed and nut alive and healthy for thousands of years. This same fulfilling protein, which is kept as a "pool" in reserve for use by the seed or nut, can be used for the same cell-feeding purpose by your body.

Combine Seed-Nut Protein with Other Protein Foods. While most seeds and nuts contain *complete* protein, some of the amino acids may be in smaller quantities, such as isoleucine and lysine. This means you would do well to eat nuts in *combination* with other protein foods such as dairy, whole grains, beans, peas and animal foods. You will then be enhancing the weak amino acids of seeds and nuts and making them more readily available for cell-tissue replication.

Four High-Protein Nuts: While all seeds and nuts are high in protein, four of the latter appear to be of higher-quality. These include peanuts, which are 25% protein, almonds, which are 18.6% protein, cashews, which are 17.2% protein and English walnuts, which are 14.8% protein. To feed your cells high protein from nuts, combine these four with an assortment of other nuts and some seeds and use regularly. It's the delicious way to feed your cells a top quality form of protein.

Seeds: Source of Life. Plants do *not* ripen to make food for us, but to encase and protect the seeds inside. We eat the delicious flavors and delicious flesh of the plants, yet it has been the seeds that made it all possible. Often, we toss away these seeds. The plant went to this effort to build the seed and stored within it the life-giving germ and a reserve of other nutrients to help it start in life again as a baby plant. Seeds are the miracle source of life for the plant. This miracle protein is now available for you in your everyday eating.

A VARIETY OF SEEDS FOR REJUVENATING PROTEIN

Seeds are sold at many large supermarkets and health stores. Here is a listing of some of the popular varieties and how their protein supplies can help give you the feeling of lasting rejuvenation.

Acorn. Ground acorn is often used as a coffee substitute. Acorns are good sources of proteins that help enrich blood cells.

Aniseeds. A prime source of those proteins that are used by enzymes for the building and rebuilding of the digestive cells and tissues to promote better absorption of ingested nutrients.

Barley. A seed food which has a balance of essential fatty acids used by protein in your body to help maintain better tissue integrity and stronger membranes.

Buckwheat. A high form of protein combined with rutin, still another ingredient, which your body uses for maintaining a more stable blood pressure level.

Caraway. A flavoring seed containing amino acids which help nourish the cells and tissues of the respiratory tract and promote better breathing.

Cardamon. An herb of the Orient, it appears to enhance the vital processes of the body. Its protein has good balance for cell-tissue nourishment and protection against cross-linkage.

Carob. The seed pods are eaten and are a source of those amino acids which repair the digestive-intestinal tract to promote healthful regularity and protection against constipation, which is a symptom of weak or broken tissues.

Celery Seed. Its amino acid pattern appears to influence the glandular network to promote better formation of hormones, needed to stimulate energy and vitality.

Coriander. A seed from the Orient that was said to promote immortality. Its high biological pattern of amino acids appears to improve the cells of the skin, giving a better appearance of youth.

Cucumber Seed. A prime source of protein that uses the minerals to help improve the health of the blood cells.

Cumin. The amino acids help promote better digestive and eliminative functions by promoting cellular repair of these systems.

Dill. A flavorful seed that is said to influence the healing of respiratory tract cells and tissues and boost better aeration of the body.

Fennel. The protein of fennel seed has a special "oil" that is used to soothe the burned tissues of the digestive tract, and bring about corrective healing.

Fig Seeds. The tiny seeds in figs, the crunchy specks that have a nice taste, should be chewed thoroughly to release the protein that can then help boost the health of the digestive-intestinal tract.

Flaxseed. The protein uses the unsaturated fatty acids of this seed to help protect against arteriosclerosis. The protein also influences the digestive-eliminative tract and promotes regularity.

Lentils. A seed-like legume that has a high protein count which repairs the iron-storing cells of the bloodstream and protects against anemia. Helps enrich the red blood cells and boost a feeling of warmth in the hands and feet.

Millet. A seed-like grain that is especially good because its protein content is high and able to use the calcium in the same seed for the building of strong bone cells.

Okra. The seeds may be roasted and ground and used as a coffee substitute. Its protein content is very high and is favorable for nourishing the body cells and tissues in a protective manner.

Parsley Seed. The protein in parsley uses the Vitamin A of the same seed to help feed the retina and cones of the eyes and thereby promote better vision.

Pinon. This is the seed of the pinon pine tree. Because it grows wild, it has not been subjected to chemicalized treatment. Its high protein is used for better cellular replication and "boosts" the sluggish function of weaker amino acids.

Pomegranate. The seeds of this fruit contain Vitamin C which works upon the protein to help create collagen, the building blocks out of which cellular "cement" can be made. Helps plump up "tired" cells beneath the skin surface.

Psyllium. The protein helps create a form of bulk in the digestive system and thereby create regularity. The protein then rebuilds the worn cells and tissues of the digestive-eliminative tract.

Rye. The seeds are tasty and have a good grain protein that helps enhance the amino acids of other foods to create better balance.

Senna. The amino acids work speedily to nourish the glands, and enter into the components of those hormones that influence emotional health and stimulate reparation of brain cells.

Sesame. The seeds are prime sources of amino acids that use the built-in supply of essential fatty acids to wash the arteries and offer protection against cellular deteriorating in the venous system.

Squash. When combined with corn, beans, fresh fruits and vegetables, squash seeds round out a high protein balance that can help nourish the body from top to toe.

Sunflower. The seeds are prime sources of almost all known elements. Its protein content tops the list, having about 52.5%. If the kernels are completely dehulled, it is a good source of most known vitamins and minerals which work harmoniously with the protein for body rebuilding. The quality of sunflower seed meal protein is exceptionally high with a biological value of 64.5%. This places sunflower seed protein in the same class as wheat and is even more desirable than soybean or corn protein.

Wheat. The seeds are almost complete protein and can promote a feeling of overall body cell-tissue repair. Chew raw wheat seeds or sprinkle over other foods, salads, stews, sauces and main dishes and improve the cell-tissue replication of your body.

A VARIETY OF NUTS FOR "PURE" PROTEIN

Nuts are available at most large supermarkets and health food stores. Select raw and non-processed nuts. If they are roasted, hydrogenated fats and salt may be used and this inhibits the amino acid absorption within the body. Buy whole nuts or shelled nuts that have not otherwise been processed or treated. Choose nuts that are clean, free from splits, cracks, stains or holes. Do not buy moldy nuts. The "meat" of shelled nuts should be plump and fairly uniform in color and size. Nuts should be part of your protein planning program for daily eating. Here is a listing of some of the popular varieties and how their protein supplies can help give you a feeling of youthful vitality.

Almond. A fine nut, rich in protein that helps create a more alkaline reaction in the system. Its protein form helps to cleanse body tissues and acts as a natural astringent.

Brazil Nut. The protein uses the essential fatty acids for helping to cleanse the cell walls and tissue membranes and protect against arteriosclerosis

Cashew. A prime source of good amino acid balance that will help the cells replicate and impart a feeling of well being.

Chestnut. The protein appears to enhance the metabolism of carbohydrates so that there is a better distribution throughout the system and a more satisfactory amino acid distribution, too.

Coconut. The protein works upon the supply of body sugar and promotes a better metabolism by strengthening the cells and tissues of the glands and respiratory tracts.

Hickory. Like the pecan, it is a good supply of high biological protein that works quickly for cell-tissue integrity.

Pecan. It has modest amounts of protein, but this nut helps metabolize essential fatty acids in the body so that it helps improve the digestive system. The protein uses the fatty acids for cellular replication of the components of the digestive network.

Pignolia or Pine Nut. It has a complete amino acid pattern that makes it favorable for rebuilding body cells and nourishing the various systems from top to toe.

Pistachio. Greenish in color, these are very high in a form of protein that nourishes the digestive cells to promote a healthfully soothing alkaline reaction. It may help guard against acidosis or acid indigestion.

Walnuts. A wide variety of walnuts makes it possible to obtain good amino acid balance. They also offer healthful oils which are used by the amino acids to moisturize the cells and tissues and maintain better well-being.

PEANUTS: NATURE'S PROTEIN PLANT

Peanuts are considered legumes or beans, but are also very much favored as a good protein food from the plant family. Peanuts contain all 8 essential amino acids in a balance that make this a needed food. Peanuts contain a variety of the B-complex vitamins and minerals that are used by protein for cell rebuilding.

You may enjoy peanuts eaten out of the shell, or have peanut butter on whole grain bread with several slices of cheese and a fruit juice as a healthful rounded out protein meal. Seek out natural or non-processed peanut butter which has *not* been prepared with hydrogenated fats and which has *not* been spiked with salt. Natural peanut butter offers a skillful Nature-created blend of high protein for better health.

How a Seed and Nut Protein Program Restored Digestive Youth.
Barbara S. was troubled with "sour stomach" and "gas distress" that
made it difficult for her to eat many meat or animal foods.
Abdominal distress following broiled liver or a steak meant that she
had to reduce their intake. But it also reduced her intake of protein.
When she was protein deficient, Barbara S. felt weak, experienced
frequent dizzy spells, her blood coloring was poor, and she saw her
face looking old beyond her years. But since she had to restrict
intake of animal protein foods, she felt there was no other choice but
to eke out her unhealthful existence. That is, not until she heard that
seed and nut foods could provide good protein in a biologically high
balance that could help make up for the deficiency of animal food
proteins. Here is how this seed and nut protein program helped
Barbara restore her digestive youth.

Simple Plan: Feature seed and nut protein in a variety of delicious
ways with your regular foods. You now *combine* seed and nut
protein with other food proteins to *complement, enhance* and
amplify any weaknesses in the amino acid patterns of those other
foods. When Barbara S. followed this 10-step Seed and Nut Protein
Program, she perked up with bubbling health. Her "sour stomach"
and "gas distress" subsided and vanished. She was more alert. Her
dizzy spells ended. She radiated the colorful bloom of youth in her
cheeks. The program was so deliciously simple, it is part of her
low-animal but high-protein plan. Here it is:

10-STEP SEED AND NUT PROTEIN PROGRAM

1. Serve yogurt or cottage cheese with assorted chopped seeds
and/or nuts; also include baked potatoes and any desired animal
protein food.

2. For waffles, biscuits or muffins, stir nuts into blended dry
ingredients before adding liquid.

3. Mix honey with peanut butter for a sandwich filling; or use
peanut butter and cheese slices for a grilled sandwich filling.

4. Add nuts to meat, poultry or seafood salads.

5. Try slivered or sliced nuts in tossed vegetable salads.

6. Use toasted coconut or toasted chopped nuts as a quick
topping for any dessert.

7. Sprinkle whole grain cupcakes with nuts before baking and
skip the sugary frosting.

8. For a protein spread on whole grain crackers or celery, add finely chopped peanuts to cream or cottage cheese that has been blended with a little milk.

9. Perk up your next vegetable souffle. Add finely chopped seeds and nuts along with the vegetable.

10. Sprinkle assorted seeds and nuts with any fresh or cooked fruit salad.

Benefits: This *combination* of different amino acid patterns provides a form of nourishment that is *complete* for the digestive system. It stimulated the increase of cell-rebuilding DNA-RNA, and promotes better protein synthesis. The *combined* proteins of seeds and nuts with other protein foods offer a protective action to guard against missynthesized or malformed cell walls that may create "cross-linkage" or broken cells, which lead to premature aging.

But *combined* proteins with seed and nut emphasis can do much to guard against this distress. This simple 10-step program promoted the catalysis of sluggish metabolism to bring about better cellular replication. Barbara S. was rejuvenated throughout her body when this *balanced protein* nourished her cells and restored the vigor of youth throughout her digestive system, the core of youthful vitality. It is the easy way to eat your way to better health and extended youth!

Seeds and nuts are self-sustaining protein foods. That is, they carry within themselves the entire source for lasting youth. Seeds and nuts use this protein for self-perpetuation. They can "live" for thousands of years when they use their own protein stores. This indicates the miracle of cell-rejuvenation made possible by good seed and nut protein foods.

IN A NUTSHELL

1. Seeds and nuts contain protein of a high biologically *complete* pattern giving them eternal life. This same protein is available to you in everyday seed and nut eating.

2. Make a selection from different types of seeds. Mix them together with other good foods for balanced protein DNA-RNA cell feeding.

3. Nuts are high in protein value and can be enjoyed in a variety of delicious ways.

4. Barbara S. recovered from "sour stomach" and "gas distress," and improved her strength, corrected her dizzy spells and boosted her coloring on a simple 10-step seed and nut protein program. She actually feasted her way to better health, thanks to protein. *Combined* protein foods hold the key to youth extension.

19

HOW PROTEIN HELPS CREATE
EXTENDED SEXUAL POTENCY

Cell therapists claim that the availability of tissue-building protein to repair and replace fragments of the body's building blocks can do more than extend the prime of life. *Cellular nutrition can help create extended sexual potency.* The availability of a *balanced amount* of amino acids can create an environment that will help nourish the cells and tissues of those body components involved with better virility in the male and more healthful sexual desires in the female. Cell therapists who specialize in nutrition for youthful vigor have suggested that a food program emphasize a balance of *all* important nutrients and a balance of *all* protein foods. The amino acids can then work harmoniously to create an extended and prolonged sexual desire for protein-ized males and females.

EIGHT "VIRILE-VITALIZER" EFFECTS OF MIRACLE PROTEIN

When balanced protein is made available to your billions of body cells and tissues, it can help promote a minimum of eight "Virile-Vitalizer" effects which include:

1. *Refreshes, Cleanses, Rejuvenates.* The accumulation of waste products of metabolism often clog up cells and tissues and interfere with sexual desires in the body. Amino acids tend to refresh body cells and tissues, cleanse away debris and promote a feeling of rejuvenation. This will help improve the metabolic process which is the impetus to spark a healthful sexual desire, capable of consummation.

2. *Regulates Body's "Sexual Clock."* Amino acids offer a built-in regulation of the body's "sexual clock" by controlling the rate of metabolism. The greater the rate of metabolism, the shorter the sexual lifespan; overactive metabolism causes "burning up" too speedily. But adequate amino acids offer a form of "health in-

208

surance" to promote a regulated metabolism that facilitates extended and prolonged potency. In effect, amino acids regulate the body's "sexual clock."

3. *Protects Against Debris Blockage.* The circulatory and glandular processes involved with a virile-vitalizer urge can often be blocked by the accumulation of debris. This is the "free radical" reaction wherein portions break off from body compounds, are highly polluted, and enter into the reactions that accelerate aging of the cells and tissues. This causes premature aging of the sexual components which consist largely of cells and tissues. Amino acids take certain nutrients and use them as an antioxidant; that is, amino acids use these nutrients to inhibit or help neutralize the sex blockage effects of debris. Amino acids use these nutrients to act as a shield against such oxidation which is the first and most significant cause of sexual weakness. Amino acids protect against this sexually depleting action.

4. *Guards Against the Mutation Aging Reaction.* Sexual weakness if manifested when molecular changes (also known as mutations) occur in broken down and damaged cells and tissues. For instance, effects from pollution, radiation from X-rays, exposure to the chemicalized environment and consequences of food pollution all cause these molecular changes which inhibit healthy sexual virility. Amino acids help repair broken and injured cells and tissues. They are available for replication of damaged organisms caused by environmental pollution, and work to repair these mutations to provide the body with vigorous cells and tissues that are needed for adequate virility in the male and female.

5. *Replaces Stress with Healthy Virility.* Stress tends to accelerate tissue breakdown and hasten premature aging of the body's cells and tissues. Stress may include illness, temperature extremes, malnutrition, emotional distress and environmental disturbances. These disrupt the natural rhythm of the biological "sex clock" and can predispose to unnecessary impotence or lost desire in both sexes. Amino acids work as building blocks to repair the cells and tissues, as a barrier against everyday stress. A well protein-ized cell-tissue body is able to cope with stress, and provide the body with the needed vigor to consummate sexual desires.

6. *Controls Cross-Linkage for Youthful Vitality.* Sexual aging is related to the cross-linkage process whereby the various molecules interconnect with others in the tissues. When this occurs, the sexual abilities decline because of this erroneous or missynthesis of cellular

walls and membranes. Adequate amino acids should be available to control cross-linkage so that cells can properly knit and provide the body with strong and youthful "virile-vitalizer" desires.

7. *Protein Acts as a Natural Aphrodisiac.* The legendary aphrodisiacs or "love potions" have their roots in protein foods. Protein exerts a decisive influence on the "synthesis" mechanism which promotes virile vitality. Basically, a deficiency in protein means that a substance, RNA polymerase, is too weak and cannot give proper instructions to another RNA messenger. This reduces the sexual urge because of a "mistranslation" of nucleic acids DNA and RNA. The result is a mismatching of needed DNA-RNA patterns. *Faulty blueprints can produce faulty buildings.* Protein helps create better blueprints for better body buildings, so to speak. This creates better "messenger" service to send "instructions" to the RNA messenger via the substance, polymerase. In effect, this provides an impetus so that *desire is fulfilled in consummation.* Protein can be considered a natural "love potion" in this body rhythm.

8. *Promotes Youthful Aeration.* Cellular aging and sexual decline are emphasized when the body's internal viscera are subjected to oxidation. (In this protein-starved process, oxygen enters a molecule and alters its function and structure. The molecule is predisposed to cellular-tissue breakdown and sexual decline.) Compare it to a dried-up rubber tube that has become oxidized. Its elasticity is gone; it has become brittle and can break under the slightest friction. Protein is the miracle substance that can promote internal ventilation so that natural aeration helps protect against oxidation. When your cells and tissues are properly aerated, they can keep your organs in good functional and working order. This is made possible by adequate protein with a balance of other essential nutrients.

Protein is Nature's miracle of cell-tissue rejuvenation to help provide the ingredients needed to nourish the components required for healthy and extended sexual potency.

COMPLETE AMINO ACIDS ARE NEEDED FOR BETTER VIGOR
OFFICIAL U.S. GOVERNMENT STATEMENT

The body must receive *all* essential amino acids so that it can promote healthy vigor. Researcher Ruth M. Leverton, [1] in *Food, Yearbook of Agriculture* tells us,

---

[1] Ruth M. Leverton, "Amino Acids" in *Food, Yearbook of Agriculture,* U.S. Department of Agriculture, Washington, D.C., 1959, page 68.

"If the intake of any one of the essential amino acids is too small to meet the body's need, none of the other essential acids being fed can be used for growth or maintenance of tissue. They will be deaminized, and the nitrogen will be excreted. (Nitrogen is the index of the amount of an amino acid involved in the body's metabolism.) Also, the body will be in negative nitrogen balance because it has to use some of its own tissue protein as a source of the needed amino acids.'

This emphasizes the importance of having *all* essential amino acids available for good nitrogen nourishment and for adequate metabolism. When *all* essential amino acids are available, they work to replenish body tissues and create wholesome health that is conducive to extended sexual potency.

Eat Combined Proteins, Says U.S. Government. Researcher Ruth M. Leverton, speaking for the U.S. Government, continues to explain,

"Our knowledge of the patterns and the amounts of amino acids in food challenges us to find ways to improve both the quality and quantity of the protein supply wherever the need exists. Sometimes this can be done by careful combinations of foods that are already available in a country."

"One way is to use a relatively small amount of a protein from an animal source to improve the quality of a protein from a plant source. Animal proteins are not always available, however, and then different plant proteins must be combined in a way to supplement each other."

"In some instances, the main cereal grain can be supplemented with another grain—such as supplementing corn with wheat to improve the amount and proportion of tryptophan. Legumes, too, may be used as supplements. To achieve the goal of a good protein supply for everyone, adjustments may have to be made in some countries in the kinds of crops grown, in the methods of food processing and preparation, and in the food habits of the people."

For extended sexual potency, obtain *all* amino acids. This will boost body health. Your ultimate reward will be a healthy and happy sexual life.

THE "NATURAL PRO-SEX POTION" FOR YOUTHFUL VIGOR

Martin B. was embarrassed by his increasing sexual weakness. Only 49, he experienced "failure" after "failure" until his frustrations made him so temperamental and angry, he became difficult to live with. His wife, understandably, shared his nervous temperament and erratic personality. She knew that he felt his "loss of manhood" as a "total loss" and embarrassment. She believed that his work as an active sales manager was draining out his energies. But even when he had a vacation, Martin B. was 'inadequate" and more frustrated than ever. Martin's wife set about to bolster his health. She knew that wear and tear from business had made him "aged" and that protein was the natural way to help restore the prime of life. This would include the *sexual* prime of life, too. She prepared this powerful protein remedy:

Natural Pro-Sex Potion. In a glass of fresh fruit juice, mix two tablespoons of Brewer's yeast, two tablespoons of wheat germ, and four tablespoons of desiccated liver (available at most health stores). Stir vigorously. Drink one glass in the morning. Drink another glass at noon. Drink a third glass in the evening.

Benefits: The vitamins and enzymes of the fruit juice are taken up by the protein in the yeast, wheat germ and liver, and they are used to speedily promote an internal regeneration of cells and tissues. Specifically, this *combination* appears to exert a decisive influence upon the components concerned with sexual consummation. The ingredients are all natural and prime sources of almost *all* known vitamins, minerals, amino acids, enzymes and essential nutrients. A powerhouse of cell-tissue replication in a tasty beverage.

Results: Martin B. took this tasty *Natural Pro-Sex Potion* three times daily. He also ate a balanced meal consisting of more fresh foods, wholesome fruits, vegetables, assorted seeds, nuts, other protein foods and less bleached or processed foods. It took *five days* before the *Natural Pro-Sex Potion* so revitalized his entire organism, that he was as energetic as a young bridegroom. Now his wife joshingly complained that if he continued feeling "so good," they would have to have separate bedrooms! Miracle protein had saved their marital happiness.

THE SEX FOOD THAT ALERTS YOUR "SEX CLOCK"

Healthy sexual desires are influenced by a *balanced protein program* containing *all* essential amino acids and many of the non-essential aminos, as well as the other needed nutrients. When your metabolism receives these amino acids in a Nature-created balance, then it can alert your "sex clock" and set it off so it can produce efficient rewards. There is *one* food that has this Nature-created balance.

That food is the *soybean.* Here you have a Nature-created balance of *all* essential amino acids, together with just about *all* other aminos, vitamins, minerals, essential fatty acids, carbohydrates, and almost all essential ingredients needed for good health. The soybean has always been hailed as Nature's all-purpose and all-inclusive food. It has been packaged this way by Nature and influences the mechanisms that wind up and alert your "sex clock" mechanism to function adequately.

The Little-Known Sexual Influence of Soybean Protein. To begin, your body's sexual responses are determined by the scientific quantum theory of energy. This law of Nature is involved with the Theory of Relativity. The formula is: $E=mc^2$. (Energy = mass times the square of the speed of light.) Every cell composing your body is a microcosm that is endowed with a particular phenomenon of matter and characteristics of its own, but it does not escape the great Nature-Science law: matter = energy = movement. Your sexual responses depend upon such energy or movement. To bring about this movement, *amino acids are needed in a balance that is provided by soybean protein. No other food in the plant world has this unique complete pattern.* Soybean protein is more effective than meat protein because it is of a plant source. It does not require the heavy metabolic processes as are so with meat, and it can be utilized *fully* by the processes needed to promote the rhythm of: matter = energy = movement.

Alerts Sexual Components into Activity. The preceding formula, $E-mc^2$, governs your sexual responses. To stimulate this process, soybean protein is used by the metabolic system to create a vibrating wave that travels along your entire organism, comparable to that of an electric current at a low voltage. You should feel every muscle, every part of your body gently vibrating with a sensory perception that is most delightful. This is the effect of Nature-balanced soybean protein.

Transforms Desire into Sexual Consummation. Irene U. may have originally felt some response toward lovemaking, but she became "cold," no matter how attentive her husband was and no matter how much he tried to "warm" up her affections. She would sob herself to restless sleep, night after night. Irene U. might have continued on in this frustration, much to the chagrin of her husband, too, except that she sought to alert herself with yoga exercises. Her yoga teacher told her, during a class session, that the Orientals prized the soybean for its unique powers of perpetuating potency. That was when Irene U. went on a surprisingly simple program which she called her "Love Bean Plan":

Daily, eat a plate of cooked soybeans mixed with natural brown rice. The weak amino acids of the soybeans are enhanced by the correspondingly stronger amino acids of the rice. This creates a combination that boosts the body's E-mc² process and winds up the "sex clock."

Together with a balanced food program of other essential nutrients, this "Love Bean Plan" helped alert her sluggish circulation, and once her body's cells and tissues were replenished, then she could respond with healthy vigor.

Happiness Is Restored, Thanks to Protein. It took four days before the "sex clock" was wound up and Irene U. could feel the warmth radiating throughout her body. Then she was a happy wife. Her husband was very happy, to put it mildly, and they felt their marriage was saved. Thanks to protein, life is now very satisfying for both of them.

PROTEIN FOODS TO NOURISH THE VIRILE NETWORK

Protein, metabolized into usable amino acids, can help nourish the millions of cells and tissues involved in the "virile network" of the body. Every component is nourished, repaired and made more youthful so that the organism can function as a whole. Some healthful protein foods that can promote replenishment of cells and tissues of the virile network include the following:

Lean Meats. A good source of all essential amino acids that help nourish the body's cells and tissues. Liver is considered a good protein food for overall cellular nourishment and better virile capability.

Seeds. Prime sources of protein factors that can nourish the prostate gland, which in turn influences male virility and fertility.

Nuts. High concentrates of protein that can replenish cells and tissues of the female ovaries and promote better production of estrogen, the hormone of youth. Protein-produced estrogen also seems to protect against skin drying and wrinkling.

Whole Grains. The good grain protein is helpful in promoting a secretion of the male hormone, testosterone, which is influential in giving him both virility and fertility. The organism involved in producing male sex liquids requires protein for activity and for manufacture.

Fish. The ocean offers a treasure of life-building protein in the form of seafood. A bonus is that the fish is a source of all known minerals from the ocean. In combination, the protein plus the minerals work to nourish the billions of body cells. They also help repair and nourish the millions of brain cells. The brain loses about 100,000 cells a day. They must be replaced because virility and potency begins with the brain and its thinking abilities. An alert protein-nourished brain can promote better lovemaking desire. Fish protein is a superior source of nourishment for the brain cells.

Eggs. Here is a complete protein food containing good-quality, complete protein with *all* essential amino acids. They do much to help nourish the millions of muscle cells and prevent them from shrinking or becoming weaker. The genitalia and pelvic area is composed of such muscles in need of complete repair and nourishment. Three eggs per week can do much to create this inner replication of essential body parts. Cholesterol watchers can enjoy just two eggs per week for good amino acid activation of essential cells and tissues.

Dairy Products. Good quality protein; combine them with grains, fruits, vegetables. Its protein is especially good for nourishing the body's glands, which influence a healthy virile response. A variety of dairy products such as milk (or skim milk), buttermilk, yogurt and cheeses of any fresh quality, together with other foods, can supercharge your glands so they can secrete needed "youth hormones," which carry more protein to those body parts concerned with consummation of the virile act.

Legumes. These include all varieties of beans, peas and foods made of them. A variety of legumes as well as sprouts will offer high concentrations of good plant protein, needed for balance, to help boost body processes for overall rejuvenation. Legumes (especially sprouts) offer high grade *net protein utilization,* or NPU, which sends essential *nitrogen* as food for the body's cells and tissues. Once they are nourished, these cells help promote better urges for biological responses.

TWO FOODS THAT OFFER COMPLETE PROTEIN FOR HEALTHFUL VIRILITY

Two foods, *combined,* offer complete protein in a favorable balance to alert the "sex clock" mechanism. These include *sunflower seeds* and *peanuts.*

Balanced Protein. Protein adds up to 25 percent of the weight of sunflower seeds. This food is low in only one amino acid, lysine. Therefore, mix sunflower seeds with peanuts which are high in lysine, and slightly lower in other amino acids. But when *combined,* these inadequacies are bolstered and you now have an amino acid pattern that is complete and balanced. This type of mixture has a good influence on your "sex clock." It's the tasty and delicious way to boost the health of your cells and tissues and supercharge your organism with vitalic living. Just mix equal portions of *sunflower seeds* and *peanuts* and use for snacks throughout the day. It'll give you good health and a wholesome capacity for good lovemaking.

SECRET OF THE "ROYAL ROMANCE FOOD" FROM THE MIDDLE EAST

The rulers of the Middle East have a traditional and legendary "Royal Romance Food" that reportedly spurs men and women onto lofty planes of romantic endeavor. It is mentioned in the Arabian classic, *Thousand And One Nights,* as revealed by Scheherazade, the wife of the sultan of India and the narrator of these tales. Scheherazade vowed that this "Royal Romance Food" could provide "unlimited capabilities" and, indeed, a "thousand and one nights" of pleasure.

Today, many still swear by the use of this simple food. The secret is that a blending of the ingredients creates a *complete* amino acid pattern that appears to exert a favorable influence upon the body. This alerts the mechanisms that stimulate the responses involved in romance. It is regarded as the most potent of all known love potions, and because it is all-natural, it is the most widely used. You can make it yourself with ingredients found in any supermarket or health store. You may have these ingredients in your pantry right now.

How to Make the "Royal Romance Food," In a glass of buttermilk (or yogurt) stir one tablespoon of honey. Now add an assortment of chopped seeds and nuts. Stir vigorously together. Then eat slowly with a spoon. Just one glass daily of this "Royal Romance Food" appears to be nourishing enough and sufficient to boost cell-tissue repair and the body components involved in vigorous lovemaking.

Benefits: Modern food scientists recognize there is more truth than romance in the effectiveness of this "Royal Romance Food." The high biological value of its *casein* is boosted and utilized by the high biological value of the *isoleucine* of the seeds and nuts. The combination then is boosted by the vitamin-mineral-enzyme potency in the two foods, and this creates a supercharging of the body's system, comparable to that of an electric charge! The ancients considered it the most vigorous of their legendary love potions or aphrodisiacs. Today, food scientists agree it is a desirable food for better health. It is a delicious way to feed yourself the capacity to enjoy longer lasting and extended sexual potency.

IMPORTANT POINTS

1. Miracle protein in everyday foods can offer a minimum of eight "virile-vitalizer" benefits.

2. For better vigor, obtain complete amino acids as recommended by the U.S. Government.

3. Martin B. conquered "failure" in marriage by a simple "Natural Pro-Sex Potion" that tasted good and restored love to his life and to that of his wife.

4. Soybean protein is a "sex food" that can alert your "sex clock" by influencing your energy generators.

5. Irene U. found "second youth" through a delicious "Love Bean Plan" calling for just two everyday foods.

6. An assortment of protein foods is available to nourish the body's virile network.

7. Combine two simple foods for complete protein to give you healthful virility.

8. Enjoy the "Royal Romance Food" made with everyday ingredients in a few moments. Enjoy cell-tissue replenishment and a "second chance" for lasting sexual vigor.

20

A TREASURY OF ALL-NATURAL PROTEIN HEALERS

Protein is the natural way to help build and rebuild your billions of body tissues. It promotes restoration of health and extended rejuvenation of your body and your mind. Here is a treasury of all-natural protein healers. Protein has an effective healing capacity when used externally, and has regenerative powers when used as part of your daily food program. Protein can be the miracle of cell-tissue rejuvenation and the key to better health when used in conjunction with balanced living and good health programs. Let protein work to give you the good health you are entitled to receive as a child of Nature. Here is a compendium of such healing remedies.

Protein-Meal-in-a-Dish

Blend cooked potatoes to equal 1 cup. Put into a bowl. Add 1 egg and stir together. Now mix in soybean flour until dough hangs together quite well, about 1/3 cup. Next, drop from tablespoon sideways into hot oil in a pan. Brown on both sides. Add chopped onion for extra flavor. Season with spices and herbs when done. This offers your body *complete* protein with all amino acids needed for cell-tissue nourishment and body health.

Strengthening Pancreas

Put 1 heaping tablespoon of bean pods into 2 cups of boiled water. Cover. Let soak overnight. Next morning, boil for 10 minutes and sip as a tea. The bean protein prepared in this form appears to be nourishing to the cells and tubules of the pancreas and offers protection against symptoms of diabetes.

Improved Respiration

Place 2 eggs in a glass or china bowl and add the juice of four lemons. Place the covered bowl in a cool place. After four days, the lemon juice vitamins will have drawn out the protein and the calcium from the eggs and made them into a form that appears to be healthful for the respiratory tract. Take 4 tablespoons of this liquid every day. Every third day, remove the eggs, discard, and replace with new eggs and new lemon juice. Respiratory cells and tissues appear invigorated with this protein-calcium mixture that is emphasized by the vitamin action of the lemon juice.

Stronger Bones

Protect against the problem of "middle age" brittle bones by using good protein and calcium. The amino acids converge with the calcium to create stronger bone marrow and protection against osteoporosis. Warm yogurt to about 160°F. Drain off the whey or liquid portion through a cloth. The firm mass which remains is a tremendous supply of good protein and calcium. It is a highly concentrated form that sends good amino acid balances to your bone structure for youthful rebuilding.

Gastric Ulcers

Rebuild the inflamed and degenerated cells and tissues of the digestive tract with a healthful protein program. Soak 3 tablespoons of whole linseed or flaxseed (available at most health stores, herbalists or pharmacies) and 6 tablespoons of whole wheat bran overnight in 1 quart of boiled water. In the morning, boil 20 minutes. Strain and let it cool. Add some honey for flavor. Sip slowly throughout the day. Amino acids from this beverage are soothing as they gently promote repair of the injured digestive tract cells.

Morning Protein Stimulator

That "tired" or "loggy" feeling may be traced to inadequate protein as body responses are hindered. Here's a *Morning Protein Stimulator* that gives you "get up and go" protein power. Two tablespoons of flaxseed, 2 tablespoons of whole wheat bran, 1 tablespoon of whole grain wheat flour, 2 sliced sun-dried figs and a handful of sun-dried raisins are combined and cooked in a cup of water for 5 minutes. Serve with some yogurt and fresh fruit slices. You nourish your body with this tremendous store of protein and complementary vitamins, minerals and enzymes. You'll feel yourself awakening with better vigor as the protein works speedily to nourish your millions of body cells and give you better physical and mental abilities.

Soothing High Blood Pressure

Amino acids are able to knit and repair the cardiovascular venous network and protect against high blood pressure. A good source of these needed amino acids is found in natural brown rice. A good *Protein Blood Pressure Tonic* can be made easily: put 3 tablespoons of natural brown rice into 1/3 quart of water. Add dried fruit slices. Boil this mixture for 20 minutes. Let cool. Eat as a soothing amino acid food to soothe the blood pressure.

Protein Sprouts for High-Powered Vitality

Your body will bounce back with amazing youthful stamina and vigor with protein sprouts. To make, just soak whole wheat grains in water at room temperature overnight. Next morning, spread out thinly on a dish or a sheet of brown paper. Do this for three days. Protect the grains from becoming sour by rinsing under free flowing cool water, three times a day. Sprouts will soon appear. When they are about ¼ inch long, they are at their peak of protein power. Sprouting can increase the protein potency of grains and seeds as much as 300%. Eat sprouts regularly for high-powered vitality.

Protein Drink

To replace the non-nutritive "coffee break," try this delicious and high-powered *Protein Drink* as recommended in the *Parker Natural Health Bulletin.*[1] Put 1 1/4 cups of milk and 1/3 cup of peanut butter in an electric blender container. Blend at high speed until smooth. Add one cup of yogurt and blend to desired consistency. (Makes three average-size servings.) *Benefits:* One serving of this *Protein Drink* supplies more than one-fourth of your daily protein needs. In addition to the protein, you receive calcium, phosphorus, riboflavin, niacin and Vitamin A. If you want to lower the fat and some calories, just substitute skim milk for the whole milk. For extra taste and nourishment, add a sliced banana. You have high-power protein to give you good "go power" for your daily activities.

Skin Feeding Protein

With a fork, mash together a raw egg, lemon juice and vegetable oil. Apply to the face. Let remain for 15 minutes. The protein helps tighten up the skin pores and nourish the cells and tissues to protect against breakdown and aging. Then splash off with alternate warm and cold water.

Better Circulation

Protein is needed to nourish the bronchial tubes and facilitate breathing. The Scotch and Welsh people live in a rough climate, yet they can cope with the altitude, and breathe healthfully, because they traditionally use oatmeal, a high-protein grain food. *To make:* place uncooked oatmeal in a glass or china

---

[1]*Parker Natural Health Bulletin,* Parker Publishing Company, Inc., West Nyack, New York 10994, Vol. 4, No. 4, February 18, 1974.

bowl. Cover with boiled water for a few minutes to soften but not cook it. Add grated apple and chopped nuts. Then eat with a spoon. You have a tremendous supply of good quality protein with vitamins and minerals that work harmoniously to feed the respiratory tract and other body parts. It gives you better circulation and improved breathing abilities.

Wrinkle Eraser

Tissue breakdown traced to inadequate protein may predispose to wrinkles. Use this simple *Protein Wrinkle Eraser:* Combine equal portions of bran and baking soda with water. When mixed into a paste, apply to your face. Add a heavier layer over wrinkles. Let remain for 30 minutes. Amino acids seep through the pores to help replace damaged tissues and "plump up" the reservoirs beneath to smooth out furrows. Rinse with lemon juice and water.

Blemishes

Mix honey with whole grain (high protein) wheat germ. Heat slightly and then spread with your fingertips over the blemishes. Let remain for 30 minutes. The protein helps draw blood to the surface through regenerated cells and to perform a natural antiseptic action on the skin surface. Wash off with warm and then cool water. Your face should feel tighter and toned up. Blemishes will gradually subside.

Blackheads

Make a paste of high-protein oatmeal, high-protein egg white and some honey. Then massage on your skin for 15 minutes. Rub over the blackheads. Let remain until it feels tight. Then splash off with tepid and cool water. Amino acids help tighten up enlarged pores and guard against infiltration of toxic elements.

Puffy, Swollen Eyes

Tissue breakdown manifests itself into puffy eyelids. Make a wet compress of six tablespoons of grated potatoes and place as a protein poultice on the eyelids and surrounding area for 30 minutes. This type of plant protein helps heal damaged cells and promotes healing and relieving of puffy, swollen eyelids.

Tightening Skin Pores

Combine almond meal with water into a paste. Spread over skin pores for 30 minutes. Then rinse and splash with witch hazel. Almond or nut protein is helpful in tightening up gaping skin pores traced to tissue breakdown.

When someone said Muriel P. was "too old" to apply for an office job, she went into action to protein-ize her body and her skin so she would pick up the slack in her face. Here is her effective *Face Pick-Up:*

Face Pick-Up

To dry oatmeal, add a whipped whole egg. Mix together, then massage into the face. Let remain up to 20 minutes. Rinse off with contrasting warm and then cool water. The protein from the whole egg (it's complete and pure protein) works with abrasive elements of the oatmeal in smoothing away furrows and nourishing the tissues so that they are plumped up and more youthful. Muriel P. eats a balanced protein diet, too. With this *Face Pick-Up,* she has smoothed out her lines, creases and wrinkles and now is "young" for any job she seeks, thanks to protein.

Protein Skin Pep Food

The so-called sallow or blotched skin color may be caused by a breakdown of the cells beneath the surface of the skin. Nourish these cells through the pores by using a *Protein Skin Pep Food:* Apply high-protein buttermilk as a mask. Let remain for 30 minutes. Body warmth keeps pores open so buttermilk protein seeps within to nourish the cells and help give them sparkling health. This helps to brighten up the discoloration of the skin and promote a more youthful color.

As an outdoor worker, Jeff B. was the victim of rough, scaly skin. Pollution robbed his skin of its protective mantle. Chemical infiltration from the environment corroded his protein stores and his cells and tissues deteriorated. This made him look older beyond his years. Rough scales gave him an unsightly "fishy" or "peeling" look. He was told of one protein program that was unbelievably simple and just as amazingly effective. Here is what Jeff B. did:

Smoothing Out Rough Scales

Every night, he would massage *mayonnaise* into his skin. Since mayonnaise is made with eggs and oils, he receives a good protein and essential fatty acid nourishment. He massaged the mayonnaise for 15 minutes, and then let it remain another 15 minutes. Then Jeff B. would splash off with contrasting warm and cold water. *Benefits:* The massage warmed the skin and opened the pores, permitting the protein from the mayonnaise to enter and nourish the damaged cells and tissues. Soon, the scales were cleared up. He radiated a youthful looking skin, and now gave the impression of being a healthy outdoor worker, thanks to his nightly protein program.

Dandruff

This is a *normal* skin function but it can be excessive if the scalp is denied needed protein. Just beat a whole egg into ordinary baby shampoo and then shampoo twice weekly. The protein of the egg will help replenish the epithelial cells and follicles of the scalp and promote better cellular replication so that dandruff (decaying or malnourished skin cells) can be controlled.

How to Feed Protein to Your Hair

Since your hair (scalp, too) is largely protein, it is important to keep it nourished with protein for good health. You can add more body to your hair by feeding it with this simple protein remedy. Make a paste of dried skim milk powder and water. Now apply to your hair as a pack for 30 minutes. The rinse off. This should help nourish your hair by giving it needed protein. A daily "protein feeding program for the hair" should help give it thicker body and youthful health.

Slimming with Protein

Calories from protein are metabolized by the body to repair cells and provide energy, and are *not* stored as fat. You can help control your appetite and slim down by eating protein foods. *Suggestion:* Put a stop gap on your appetite by eating two tablespoons of natural peanut butter. It helps fill up your "empty feeling," and will later enable you to eat much less and keep yourself slim, while feeding yourself protein.

Soothing Swollen Feet

If your feet are puffy from too much walking or standing, bathe them in a solution of protein. Just four tablespoons of skim milk powder to a quart of warm water. As your foot pores open, the protein enters to replenish broken cells and tissues, giving your feet a feeling of "glad all over."

Stronger Finger and Toe Nails

Since the nails, themselves, are large forms of hard protein, a body deficiency can be symptomatic in splitting or brittle nails. Boost protein intake through better assimilation. You can mix one tablespoon of unflavored gelatin in a glass of fruit juice. Then drink up to three such glasses daily. The protein of the gelatin is boosted by the enzymes and Vitamin C of the fruit juice, and sent to all extremities, such as your fingertips and toetips, for nourishment. This will help improve the strength of your finger and toe nails.

Protein Bath

Refresh yourself all over by soaking your body in a Protein Bath. To prepare, mix 1 cup powdered milk with 4 tablespoons vegetable oil. (Wheat germ oil is best since it has more protein.) Add ½ cup table salt. Mix together into a paste. Then add to a quarter-filled tub of comfortably warm water. Swirl vigorously with your hand to disperse the mixture. Then just soak yourself in the bath. Your body pores open and welcome the warm infusion of protein. After 30 minutes, you'll emerge feeling regenerated, thanks to the potent protein replenishment of your body cells and tissues. You'll look good, too.

Knee and Elbow Softener

Mix 1 tablespoon of whole grain flour with 1 tablespoon of ordinary cold cream. Work this paste into your knees and elbows. Let remain up to 30 minutes. Then splash off with a warm-water washcloth. Repeat daily. The protein from the whole grain flour will help slough off decayed cells, and will then nourish the "starved" knees and elbows to keep them fresh and youthful with new cells and tissues.

Lips and Mouth

Help control and erase unsightly wrinkles caused by protein deficiency and cellular breakdown by applying wheat germ oil to the lips and mouth. Let remain overnight. A few nights should help replenish decayed cells with protein from the wheat germ, and should reward you with smoothly youthful lips and mouth skin.

Chapped Lips

Just apply any whole grain oil such as soybean oil, sunflower seed oil, sesame seed oil or wheat germ oil to the lips. The protein joins with the essential fatty acids to moisturize the cells of the lips so they can "drink" and correct their parched dryness.

If you look and feel tired because of a strenuous day, but need to look and feel youthful because you're going out at night, then replenish lost protein with a simple refresher.

Protein Refresher

Just mix egg white (pure protein) with honey, then coat your face and neck with the mixture. Let it dry for 15 minutes. The protein does much to pick up your sagging skin cells and replenish your tissues. Then rinse with warm and cold water splashes. You'll feel refreshed.

Skin Sags and Furrows

Tighten up these protein-starved reactions with a protein masque. Mix together 1 tablespoon of honey with 4 tablespoons of sour cream into a paste. Apply to your face. Let remain up to 30 minutes. The protein from the sour cream goes deep into your furrows, and replenishes cells and tissues to help correct the wrinkling. After 30 minutes, splash off with warm and then cold water. A daily protein masque is healthful food and natural protection against cell-starved furrows.

*   *   *   *

In any form, balanced protein is Nature's "staff of life." It is the key to helping your cells and tissues become nourished and re-generated. It is the miracle all-natural food that is available to everyone in a variety of different forms and shapes. Miracle protein is the youth-restoring element of which your very core is made. It has helped thousands and now offers you hope for a longer and more youthful life. It is ready for your use. The next step is up to you!

PROTEIN PLANNER INDEX